I0090096

Copyright © 2025 Chris DeArmitt

All rights reserved. This book or any portion thereof may not be reproduced or used in any manner whatsoever without the express written permission of the publisher, except for the use of brief quotations in a book review.

First Printing, 2025
ISBN 978-0-9978499-9-8
Phantom Plastics LLC
Ohio, USA

Disclaimer
The purpose of this book is to inform by providing peer-reviewed scientific evidence, based on what was known at the time it was written. The author and publisher have made every effort to ensure that the information in this book was correct at the time of writing. Topics are included to highlight issues and are not intended to portray any company or individual in a negative light. The author does not assume and hereby disclaims any liability to any party for any loss, damage, or disruption caused by errors or omissions, whether such errors or omissions result from accident, negligence, or any other cause. If you do not agree to be bound by this disclaimer, please do not continue reading.

ENDORSEMENTS

"Chris's 'Shattering The Plastics Illusion – Exposing Environmental Myths' is a compelling and meticulously researched book that debunks common myths about plastics. Drawing from over 4000 peer-reviewed studies, Chris presents a fact-based narrative that is both accessible and enlightening. His dedication to scientific accuracy and independence makes this book an essential read for anyone interested in the environmental impact of plastics."

Joshua Otaigbe
Professor Emeritus, The University of Southern Mississippi & Principal Partner, Flaney Associates LLC, Hattiesburg, USA

"In a world of 'experts' losing trust because of their self-interested biases, whether ideological or financial, the intersection of public policy and science desperately needs truly independent research and analysis. Chris DeArmitt (himself a PhD chemist) provides exactly that when discussing the surprisingly controversial topic of plastics."

"Plastics, like carbon dioxide, 'global warming', fluoride, peanut butter, and so many other things these days are often decried as not just risky but existentially dangerous. That is, of course, because there are no clicks to be had, no grants to be gathered and no donations to be grifted by saying, 'This isn't going to kill you. Everything's just fine.' Saying such things requires persistence, courage, and even a willingness to be ostracised by those 'scientists' whose cushy, albeit mendacious, existence is threatened by the truth. Those, like Dr DeArmitt, who have such courage should be lauded and promoted."

"Chris DeArmitt discusses plastics from the perspective of a clear-eyed scientist, from biodegradation, to (now fashionable fear of) microplastics in the ocean and in human bodies, to comparing plastic PEX home water pipes to copper, to ways to manage plastic waste, DeArmitt's approach is simultaneously analytical yet enjoyable, data-driven yet conversational. 'Shattering the Plastics Illusion' is exactly what the public policy discussion around plastics needs so that politicians, regulators, and citizens work from a basis of truth rather than being misled by activists."

Ross Kaminski
Host, The Ross Kaminsky Show
Weekdays 9 AM to Noon MT
KOA 850 AM and 94.1 FM (Denver)

"All materials have some effect on the environment. This is also true for polymers. However, there are many 'fake' statements/news about the negative environmental impacts of polymers. This volume — Shattering the Plastics Illusion — by Chris DeArmitt presents an exposé about what is incorrectly stated about polymers and what is actually true. As retired but still active professor of polymer science, I will definitely use DeArmitt's book in my consulting contacts with companies and authorities."

"Polymers (plastics, rubbers and fibres) are necessary for a sustainable society. The distribution of electric energy requires cables which are insulated with plastics or rubbers. Plastic pipes are important for a sustainable infrastructure. Both are good examples of long-term infrastructure with expected lifetimes of the order of 100 years. Fibres are used in composites in order to obtain materials of high stiffness and strength with the import added value of low mass. Lifetimes of these products, important to a sustainable society, is in the range 10 to 100 years. Rubber materials with their unique properties — elasticity, high maximum elongation and in many cases high durability — are important for sustainability. Polymer packaging materials makes the food much more available by the protection towards degradation. They are also of low mass and have a much lower global warming effect than metals and ceramic materials in the food chain."

Ulf W. Gedde
Professor Emeritus, KTH Royal Institute of Technology, Sweden

"I have known Chris for many years and always been impressed by his logical and evidence based approach to important issues. In this book he has been able to crystallise many of the concerns that I have about the reporting of the plastics issue, but been too daunted by the system to try to do something about. He provides well researched independent evidence for his statements and also suggests ways to remedy the situation."

Professor Roger Rothon
Manchester Metropolitan University, UK

"Plastics are a very useful and indispensable material, but this is often overshadowed by blanket criticism and prejudices. That's why it's important to deal with them in depth, and I'm grateful to Chris for doing so with his new book."

Markus Steilemann
Chairman of the Board of Management, Chief Executive Officer at Covestro

"When we look at the facts, we can make the best choices on an individual as well as societal level. Chris helps us through his book and his incredibly analytical approach to unearth the facts around plastics. The outcome is sometimes quite surprising."

Philipp Lehner
CEO, ALPLA Group

"Dr. Chris DeArmitt's Shattering the Plastics Illusion — Exposing Environmental Myths is a bold and much-needed voice in the global conversation about sustainability. The book is based on concrete scientific studies and facts, which helps the author to clarify the numerous myths concerning plastics and provide the readers with the real solutions that are based on the facts instead of fears.

Dr. DeArmitt shows how plastics help solve problems of waste, emissions, and quality of life, while inviting us to change our thinking. This book is a must-read for policy makers, business leaders and anyone who wants to create a sustainable future, not just based on opinions but on facts.

A powerful and enlightening guide that I wholeheartedly recommend."

Carlo Bergamaschi
Executive Director, Valgroup

"They say the second movie in a series never surpasses the first. Well, that might be true for some films, but it certainly doesn't apply to the second edition of Chris's book. As I turned the first pages, I could almost hear Chris's voice, with its blend of English, Swedish, and American accents, passionately expressing his message. What makes the content of this book so powerful is the love and dedication of someone who seeks the truth based on facts, driven by a strong purpose to debunk the false narratives echoed across media and social networks.

"Chris is not only committed to his two daughters but also to creating a legacy that inspires over 8 billion people to care for this planet responsibly. As he often points out, a plastic bottle doesn't end up in the ocean on its own, just as no one would leave a plastic banknote lying on the beach. It's our shared responsibility to recycle and recognise that, among all packaging options, plastic—if properly recycled—would become the only sustainable powerhouse in packaging for the planet.

"If everyone understood the value behind a plastic bottle—from its production, transformation, and distribution to its collection, recycling, and remanufacturing—and the impact it generates in terms of jobs and the economy, they would never leave it on the ground, just as they wouldn't leave a banknote behind. This book is an invitation to understand and appreciate this journey. Read it, share it, and be part of this movement of truth and saving our planet!"

Evandro Pereira
Managing Director, South America at Plastipak

PREFACE

The book *The Plastics Paradox* came into existence because my two daughters were taught misinformation at school and that made my blood boil. As a scientist and professional problem-solver, I know that the only way to solve problems is to start with facts. Attempting to solve problems based on missing or faulty information doesn't work — in fact, it often makes matters worse.

In 2019, when I began fact-checking what we've been told about plastics and their effects on the environment, I forced myself to read over 400 peer-reviewed studies. Why so many? Well, this is a complex topic, and one needs to understand materials use, waste, litter, ocean plastics, degradation, microplastics, toxicity, and all the related topics in order to develop a full understanding.

The layperson bases their opinions on internet gossip and headlines, but the professional scientist must check all the evidence first and only then come to a conclusion. That's a huge amount of work, which may be why no one else thought to do it. The other reason is funding. Creating *The Plastics Paradox* cost me hundreds of hours of unpaid time and thousands of my own dollars.

I intensely dislike writing books precisely because it is so much work, and unless you know Oprah or have the following of a J. K. Rowling, your books are very unlikely to ever be read, so your effort is likely wasted. Once I finished the book, I relaxed, knowing that I had done my duty as a scientist, and felt safe knowing that, like my first book, hardly anyone would ever hear about it, let alone read it.

I was wrong.

Readers began reaching out. They loved the book and asked about translating it into other languages. Exhausted, I declined. But then, something incredible happened: People began translating it themselves for free. Volunteers reformatted the text for new editions, and soon the book was available in English, German, French, Italian, Portuguese, and Spanish. To those generous individuals, I owe my deepest gratitude.

You might think this led to some kind of financial windfall. It didn't. The book was offered as a free download — no email sign-up, no strings attached. Companies asked permission to print tens of thousands of copies to distribute freely, and I agreed. They sent copies to journalists, clients, and even politicians. One company mailed 535 signed copies, one for every member of the US Congress. Another sent signed editions to the Canadian Parliament. On 28 November 2024, MP

Lianne Rood quoted the book in the Canadian Parliament in a discussion about amending the Canadian Environmental Protection Act, 1999, using *The Plastics Paradox* to demonstrate why plastic bans are unwise.

The book's reach extended far beyond North America. Many thousands of copies were distributed across Europe and South America, with significant uptake in countries like Germany, Italy, France, and Brazil. As the message spread, so did invitations for podcast interviews, newspaper articles, radio spots, and TV appearances — not only national but even international.

Meanwhile, I kept reading. Over the past 5 years, I've spent thousands of hours unpaid poring over 5,000 peer-reviewed studies. Independence was crucial; no sponsors or hidden agendas influenced my work. Today, I'm recognised as the leading independent expert on plastics and the environment — not because I'm the smartest, but because I was foolhardy enough to put in the time and effort.

The message has gained traction, and I'm now invited to keynote conferences around the world. Despite all the travel and attention, my message remains simple: Facts lead to better futures.

So, why am I sitting at a keyboard again, writing yet another book? Am I a glutton for punishment? Maybe. But there's a deeper reason. While *The Plastics Paradox* remains accurate, my understanding has grown. I've read 10 times more science since its publication, and my views have evolved.

This new book, Shattering the Plastics Illusion, aims to refine and expand what we know. In addition to summarising the latest facts, it offers a comprehensive perspective and actionable solutions for a brighter, more sustainable future.

Studies will be quoted verbatim "*in blue italics like this*" for maximum accuracy.

Let's dive in…

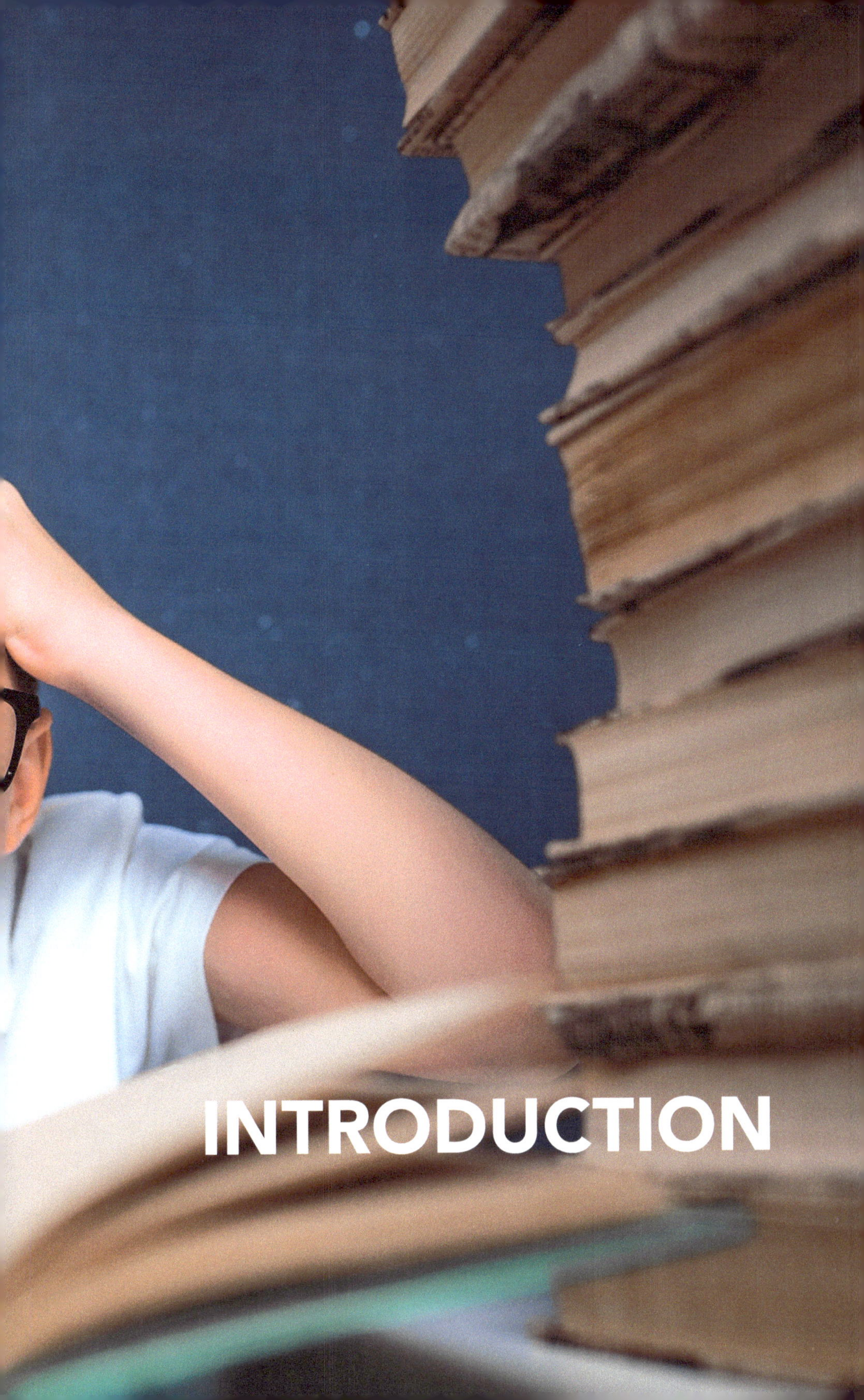

INTRODUCTION

INTRODUCTION

We all "know" a lot about plastic from the mainstream media and the internet. But here's the twist: Confidence in those sources is at an all-time low. In other words, many of our beliefs about plastics — and countless other topics — come from information sources we've openly acknowledged as untrustworthy.

"Americans continue to register record-low trust in the mass media."

"For the third consecutive year, more U.S. adults have no trust at all in the media (36%) than trust it a great deal or fair amount. Another 33% of Americans express 'not very much' confidence."

Americans' Trust in Media Remains at Trend Low, Gallup, October 14th 2024

Makes you think, doesn't it? People form strong convictions on important issues — even when they admit that their information sources are flawed. This phenomenon has a name: the Gell-Mann amnesia effect.

Worse, lies tend to stick when repeated enough, regardless of how smart you are. That's the illusory truth effect at work.

"In line with previous work, we found individuals tend to believe repeated information more compared to new information."

"Across seven studies, this tendency was not reliably and substantially related to cognitive ability..."

J. De keersmaecker et al., Investigating the robustness of the illusory truth effect across individual differences in cognitive ability, need for cognitive closure, and cognitive style, Personality and Social Psychology Bulletin, 46 (2), pp. 204–215, June 2019

What is the truth, anyway? For the purposes of this book, let's define it as "that which is backed by the most solid evidence." Everything you'll read here will be supported by evidence, with citations provided so you can verify it yourself. The facts in this book aren't my opinions — they're drawn from decades of peer-reviewed science. While others may twist the truth for profit, my goal is to give you the facts for free.

Here's another critical piece of the puzzle: Negative news dominates, not because it's accurate but because it's effective. Studies show that bad news grabs our attention more than good news, which is why it's everywhere. This media bias reinforces false narratives and skews our understanding of reality.

"Data from four US and UK news sites (95,282 articles) and two social media platforms (579,182,075 posts on Facebook and Twitter, now X) show social media users are 1.91 times more likely to share links to negative news articles."

"Additionally, the heightened sharing of negative articles to social media may incentivise journalists to write more negatively, potentially resulting in increased negative news exposure even for individuals who rely solely on online news sites."

J. Watson et al., Negative online news articles are shared more to social media, Nature — Scientific Reports, 14, 21592, 2024

This highlights the importance of rejecting false, sensational narratives. No matter how dramatic or exciting they may seem, genuine progress comes from relying on accurate, neutral, and vetted information from credible scientists.

Now, let's take a closer look at some accusations levelled against plastic.

- We're drowning in plastic.
- Plastic is filling up our landfills.
- Plastic pollution is everywhere.
- Plastics use too much oil.
- Plastics are bad because they are made of fossil fuel.
- Plastics increase greenhouse gas.
- We should switch to greener alternatives.
- The oceans are clogged up with plastic.
- Plastic harms turtles and whales.
- There will be more plastic than fish in the oceans by 2050.
- Plastics take 400 or 1000 years to degrade.
- Plastics are toxic.
- Plastics leach harmful chemicals.
- We eat a credit card of microplastic every week.

These claims have been repeated so often that they're accepted as truth by the public, teachers, journalists, and even policymakers. But are they accurate? What does the evidence actually say? The rest of this book will put these statements under the microscope, comparing them to what scientists have discovered.

Everyone has an agenda, so let me lay mine out clearly. I am a crusader for truth. Why? Because finding the facts — and then acting wisely based on them — is the only reliable path to real progress. It infuriates me that people lie to us, manipulate us, and exploit our good intentions for their own personal gain.

As a scientist, my goal is simple: to provide you with accurate, unbiased information so you can make informed decisions. Whether you choose paper, metal, glass, wood, cotton, silk, or plastic is entirely up to you. It doesn't matter to me — what matters is that your choices are based on facts, not deception.

MATERIALS IN
PERSPECTIVE

MATERIALS IN PERSPECTIVE

There is a perception that we are "drowning in plastic." So, how much plastic do we use every year relative to other materials? That is information I only discovered after finishing *The Plastics Paradox*. I was reading a book by Michael Ashby, and I turned over the page to see a pie chart showing that concrete, metal, and wood make up approximately 99% of the materials we use by weight. The number shocked me. In fact, I was so surprised that I had to check the claim against other sources.

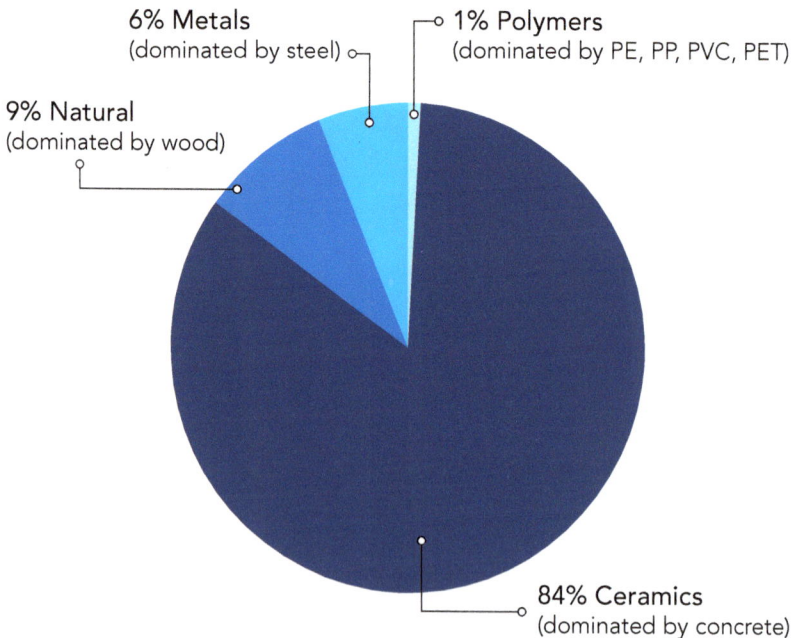

6% Metals
(dominated by steel)

1% Polymers
(dominated by PE, PP, PVC, PET)

9% Natural
(dominated by wood)

84% Ceramics
(dominated by concrete)

Materials and the Environment: Eco-Informed Material Choice, Michael F. Ashby, Butterworth-Heinemann / Elsevier, Oxford, page 18, UK, 2009

Plastic consumption is around 0.4 billion tons per year, so let's put it into perspective by comparing that to the total amount of materials we consume per year, which is 107 billion metric tons per year. A quick calculation reveals that plastics make up less than 1% of materials we use by weight or by volume. So, although we do indeed use a lot of plastic, no rational person could make the claim that plastics are the major problem when other materials account for over 99% of the total.

H. Bruyninckx: Global Resource Outlook 2024: Bend the Trend, UNEP, p. 26, 2024

Annual production of plastics worldwide from 1950 to 2023, Published by Statista Research Department, Nov 21st, 2024

We can already see that misleading information grossly misrepresented the relative contribution of plastic compared to other materials.

When confronted with the actual amount of plastic relative to other materials, some people respond that it simply can't be right because plastics are all around us. I decided to check what materials are used to make a house to gain some insight and perspective.

These are just rough numbers for illustrative purposes…

House Structure: A typical single-family, two-storey house with a wood frame and brick exterior weighs around 100,000 to 200,000 pounds (50,000 to 100,000 kg).

Furnishings & Appliances: Around 8,000 to 12,000 pounds (3,600 to 5,400 kg).

Personal Belongings: Personal belongings around 1,000 to 2,000 pounds (450 to 900 kg) per person living in the house.

That works out to 100 parts house structure to 4 parts furnishings and around 1–2 parts belongings. We overlook the materials comprising our homes in our daily lives; they are nearly invisible to our conscious minds. We also think very little about our furniture or appliances. What we primarily focus on is our personal belongings, as we feel more of a connection to them and interact with them more physically through touch, smell, and so on.

The same applies to what kinds of materials are used to make the house. Again, here are some rough estimates only to illustrate the concept. The breakdown mimics closely the numbers we just saw for total materials use globally.

Concrete: 60–70%
Brick: 5–15%
Wood: 10–15%
Glass: around 5%
Steel: around 5%
Other: around 5%
Plastic: 1–3%

It appears then that our awareness of and focus on plastic materials is very much out of proportion to how much of it we really use.

PLASTIC MARKET GROWTH

Another common claim made against plastics is that they are bad because they are "growing exponentially." So-called "environmental groups" say that all the time. Does that argument hold water?

Here is the data on how the consumption of materials has grown over the decades.

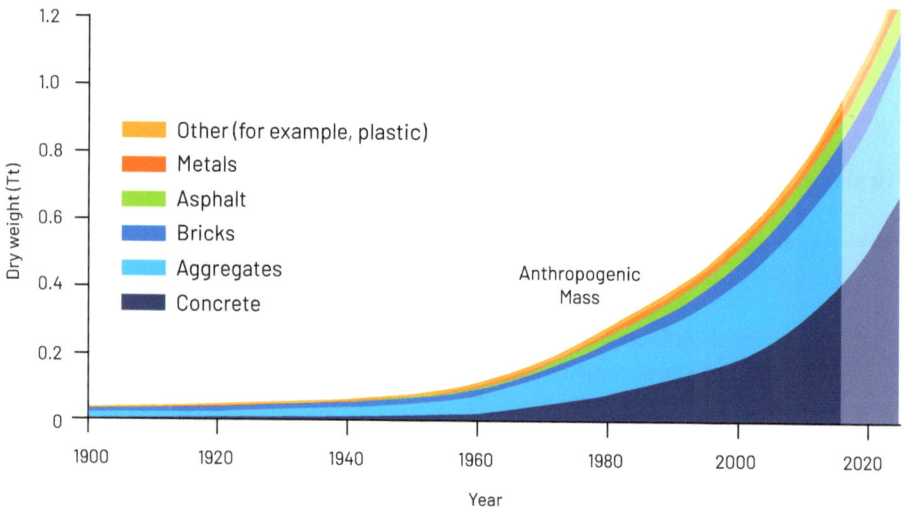

E. Elhacham et al., Global human-made mass exceeds all living biomass, Nature, Vol 588, pp 442, December 2020

It is clear from this that all materials are growing exponentially. In fact, plastic materials make up a tiny fraction — less than one tenth — of that yellow line labelled "other." Therefore, any organisation claiming that plastics are especially problematic because of their growth rate is not being honest. Plastic production has increased and is increasing in line with all the other materials we use.

Using more materials is associated with wealth. In fact, there is a linear correlation between material usage and prosperity. This means that using more materials is not necessarily bad.

T. Gutowski et al., Why We Use More Materials, Philosophical Transactions A, The Royal Society, 375, 20160368, 2017

MATERIALS & CARBON DIOXIDE (GHG)

Of course, the amount of material we use is only one factor. What if plastics are vastly worse for the environment compared to these other materials that we use more? That is a topic explored in depth later on, but let us take a first glance at it here. Impact takes many forms, but most consider that carbon dioxide, i.e. one of the greenhouse gases (GHG), is the main one. I would like to mention that I am not making any statements about global warming here; rather, I am showing the data because so many people believe strongly that this is a major area of concern.

Here is a comprehensive breakdown of GHG sources globally.

Global greenhouse gas emissions by sector

This is shown for the year 2016 — global greenhouse gas emissions were 49.4 billion tonnes CO_2eq.

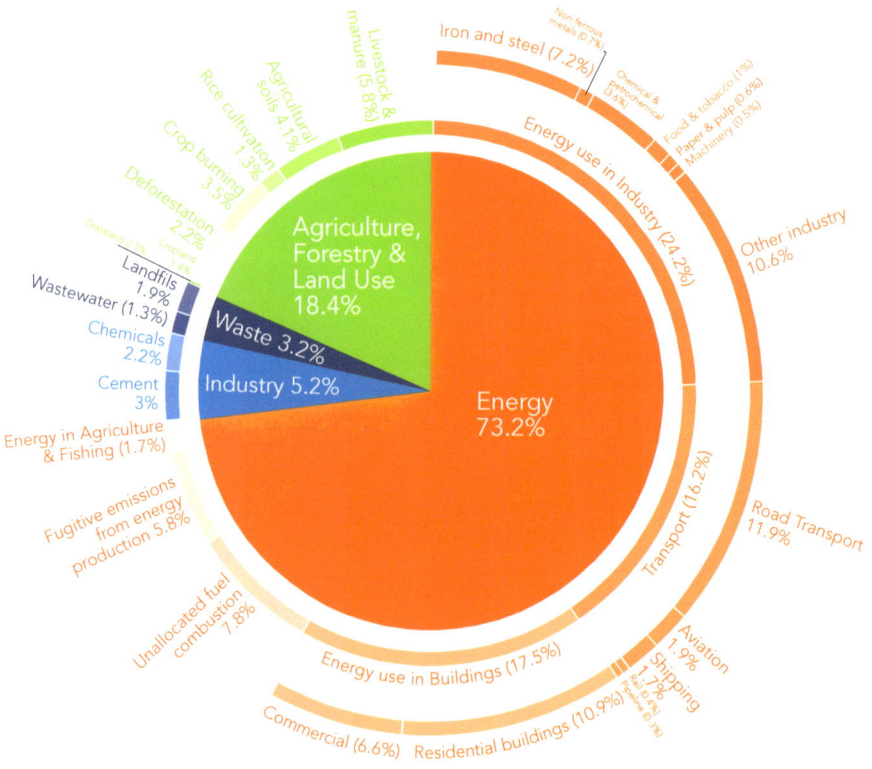

OurWorldinData.org – Research and data to make progress against the world's largest problems.
Source: Climate Watch, the Worl Resources Institute (2020).

Licensed under CC-BY by the author Hannah Ritchie (2020).

Materials production accounts for 20–25 % of global GHG emissions.

E. G. Hertwich, Increased carbon footprint of materials production driven by rise in investments, Nature Geoscience, 14, pp. 151–155, 2021

That means that materials production is clearly a major GHG factor, but are plastics the primary culprit? Industrial emissions from materials production total 10 Gt of carbon dioxide, and here is a breakdown by material.

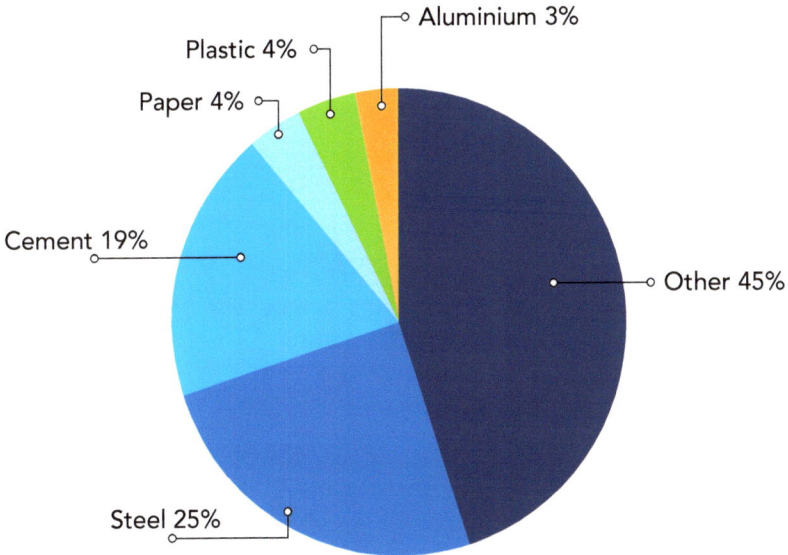

Plastic 4%
Paper 4%
Aluminium 3%
Cement 19%
Other 45%
Steel 25%

J. M. Allwood & J. M. Cullen, Sustainable Materials: With Both Eyes Open, Cambridge University Press, 2018 -From International Energy Agency Data

The data makes one thing clear: Anyone truly concerned about the impact of materials on greenhouse gas emissions should focus on iron, steel, and cement — not on plastics.

Moreover, numerous studies show that plastics can actually help reduce carbon dioxide emissions. For instance, plastic packaging plays a critical role in preventing food waste by extending shelf life and protecting food from damage. Since food production is a major contributor to greenhouse gas emissions, reducing food waste through effective packaging has a significant, positive environmental impact.

"In 2007 the estimated use benefits were 5-9 times higher than the emissions from the production and recovery phases."

"In 2020 the estimated use-benefits could be 9-15 times higher than the forecast emissions."

"Substitution of plastic products by other materials will in most cases increase the consumption of energy and the emission of greenhouse gases."

H. Pilz, B. Brandt, and R. Fehringer, The impact of plastics on life cycle energy consumption and greenhouse gas emissions in Europe, denkstatt GmbH, 2010

They discovered that plastic packaging prevents far more GHG emissions than its production generates. This illustrates the importance of considering all factors when assessing impact.

I discuss that subject in more detail later in the book.

PLASTICS ARE MADE OF FOSSIL FUEL

The common perception is that plastics are bad because they are made of fossil fuel. But does that argument stand up to scrutiny? The graph below went viral when I posted it online, garnering well over a quarter million views.

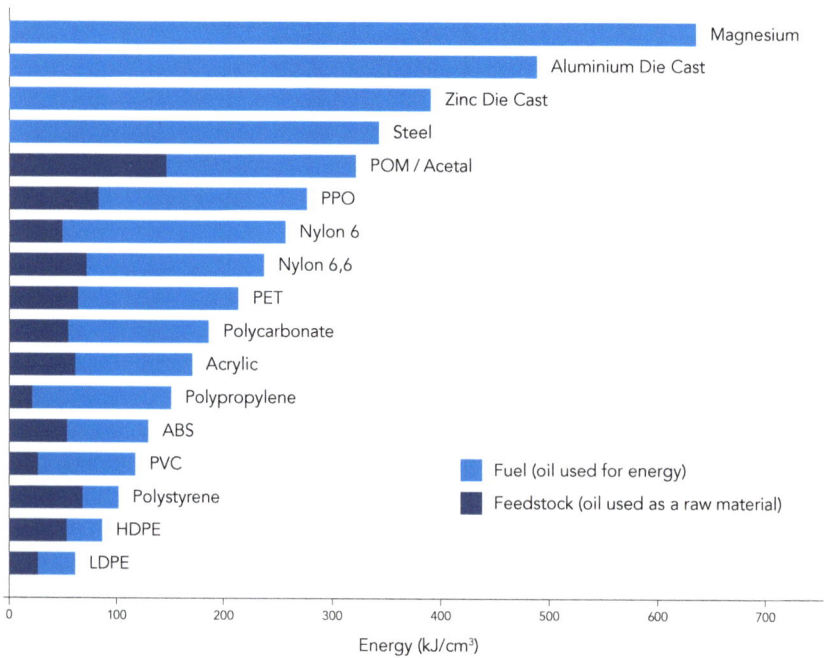

Fuel (oil used for energy)
Feedstock (oil used as a raw material)

Energy (kJ/cm³)

N. G. McCrum. C. P. Buckley & C. B. Bucknall, Principles of Polymer Engineering, Oxford University Press, UK, 1988

The dark blue bars represent the oil used to produce plastic materials, while the light blue bars show the oil burned for energy during the production process. Interestingly, the amount of oil required to make plastics is significantly lower than for many other common materials. This is because plastics are created and processed at relatively low temperatures.

In contrast, materials like iron, steel, copper, aluminium, silver, gold, and glass are processed at temperatures so high that they glow red-hot. Even without diving into the exact numbers, it's easy to understand just how energy-intensive these processes are compared to plastics.

N. G. McCrum. C. P. Buckley & C. B. Bucknall, Principles of Polymer Engineering, Oxford University Press, UK, 1988

PLASTICS USE FOSSIL FUEL

We all know that fossil fuel is used to make plastics, and that is one of the major criticisms of plastic. While it is true that oil is the raw material for plastic, there is a lot more to the story than that.

As this chapter is about perspective, we should first look at what oil is used for. Here is a depiction of the fate of an oil barrel.

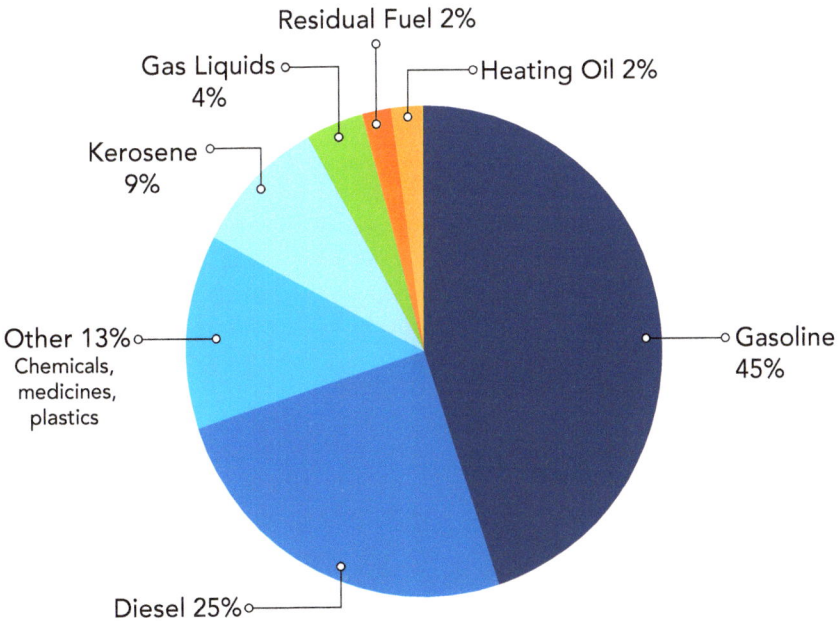

https://www.breakthroughfuel.com/blog/crude-oil-barrel
https://elements.visualcapitalist.com/visualizing-the-products-and-fuels-made-from-crude-oil/

It turns out that around 85% of oil is burnt, and the strange thing is that no one seems to mind. We jump in our cars and drive around, sometimes just for the fun of it. We get our orders delivered by truck. We jet around for our holidays. We heat our homes with oil. And yet, when it comes to using around 5% of oil to make plastic, now suddenly people protest — This must stop!

The 13% of oil use marked "other" includes chemicals, medicines, and plastic. The best scientists I know believe that we should stop wasting the oil we burn (around 85% of the total oil), instead saving it for those other far more valuable uses. How can people honestly think that burn-ing oil, which converts it into CO_2, a greenhouse gas, is a better idea than making plastic products that improve our lives and save lives? And anyway, when a product comes to the end of its life, we can still burn the plastic and release the energy to create electricity.

There is another overlooked factor when it comes to oil and plastics. So-called "environmental" groups like to talk about the oil used to make plastics, but they never talk about the other side of the equation. What does that mean? Well, we all know that plastics make cars and planes lighter, which lowers fuel, i.e. oil, con-sumption. Similarly, plastics are used to insulate buildings, so we need less energy (less fossil fuel) to heat our homes. In order to accurately and fairly assess the impact of plastics on fossil fuel and oil use, we would need to factor in the oil saved by using plastic. That is a little outside my field, so I asked 4 independent scientists to run the calculation, and the amount of oil saved by making cars and airplanes lighter may be more than the total amount of oil used to make plastic. You can even check it for yourself on ChatGPT to get a rough idea.

When you run the numbers, plastics make cars around 10% more fuel efficient. Now look at the amount of gasoline used for cars in the graphic: 43% of total use. If you save 10% of that 43% by making cars lighter with plastic, then you have saved around 4% of all oil consumed worldwide, which is about the same amount used to create all plastic materials. Adding the oil saved from lighter trucks, planes, and building insulation, plus reduced food waste, the plastics industry becomes net fossil fuel neutral or even negative, saving more oil than it consumes.

J. Allwood & J. Cullen, Sustainable Materials - with both eyes open: Future buildings, vehicles, products and equipment - made efficiently and made with less new material (without the hot air), UIT Cambridge Ltd, 2012

"Trucost estimates that if plastic components in passenger vehicles produced in North America in 2015 were replaced with alternative materials, the vehicles would require an additional 336 million liters of gasoline and diesel to operate over their lifetimes. The environmental

cost of producing, distributing, and combusting this fuel in the first year is estimated to be US$176 million and US$2.3 billion over the lifetime operating mileage of vehicles produced in 2015. This equates to an environmental cost of $169 per gasoline or diesel passenger car sold in North America in 2015."

R. Lord, Plastics and Sustainability: A Valuation of Environmental Benefits, Costs and Opportunities for Continuous Improvement, Trucost, 2016

This shows just how important it is to look at both sides of the equation. Anyone who talks only about the oil used to make plastic while conveniently "forgetting" to mention the oil saved by using plastic is gravely misleading you.

BIO-BASED PLASTICS

There are many types of plastic that can be made from plant-based feedstocks instead of oil and other fossil fuels, but the ones with the most promise are standard PE, PP, nylons, and PET made from plant-based raw materials. These are drop-in alternatives to fossil-fuel-derived plastics that also benefit the economy while having a low impact on the environment.

"The US$87 million investment aims to meet the growing global demand for sustainable products. The plant now operates at an increased capacity, from 200,000 to 260,000 tons/year."

"Braskem's bio-based ethylene is made from sustainably sourced, sugarcane-based ethanol which removes CO_2 from the atmosphere and stores it in products for daily use."

"Each ton of plastic resin made from renewable feedstock represents the removal of 3 tons of CO_2 from the atmosphere. Since the plant's beginning in 2010, more than 1.2 million tons of I'm green™ bio-based polyethylene has been produced. The recent increase in production capacity will remove approximately 185,000 tons of CO_2 equivalent per year."

https://www.braskem.com.br/imgreen/details-news/braskem-expands-its-biopolymer-production-by-30-following-an-investment-of-us-87-million

The public is unaware that plastics can and are made using other feedstocks and that we already have options to reduce reliance on oil when we need to. For the moment, it makes the most sense to reduce the burning of fossil fuels and reserve them for more valuable uses like making medicines, plastics, and chemicals. Later, we can transition to plant-based feedstocks if necessary.

There are several other plastics that can be made from bio-based, renewable feedstocks such as PLA and PHB/PHA, but life cycle studies show that they have a greater environmental impact than standard plastics such as PE and PP.

M. Tabone et al., Sustainability Metrics: Life Cycle Assessment and Green Design in Polymers, Environmental Science & Technology, 44 (21), pp. 8264–8269, 2010

PLASTIC WASTE

What about plastic waste? Is it really the fundamental problem for waste generation and landfills? From the previous discussion, we know that plastics make up less than 1% of the materials we use, so it should come as no surprise that plastics are also under 1% of total waste.

It is difficult to get an exact number for the amount of waste generated globally because 97% of waste is industrial waste and it is not documented as well as one might hope. However, various estimates reveal that 97% of all waste is industrial, and plastic makes up a minuscule portion because a large amount comes from sources like mining waste.

Elizabeth Royte, Garbage Land: On the Secret Trail of Trash, Little, Brown and Company, 2016

US Congress, Office of Technology Assessment - Managing Industrial Solid Wastes from Manufacturing, Mining, Oil, and Gas Production, and Utility Coal Combustion, OTA Report No. OTA-BP-O-82. Washington, D.C.: US Government Printing Office, 1992

Human Activity and the Environment, Minister of Industry, Government of Canada, Statistics Canada, 2012

M. Liboiron, Municipal versus Industrial Waste: Questioning the 3-97 ratio, Discard Studies, 2016

We constantly hear that plastic makes up a large proportion of waste — for example, a US Environmental Protection Agency figure stated it was around 13%, as mentioned in *The Plastics Paradox*. What I did not realise back then was that although plastic is around 13–15% of household waste, household waste is only about 3% of total waste, with industrial waste making up the other 97%, as previously mentioned.

So, far from being the major contributor to waste, as we are told, other materials make up over 99% of our waste problem. That means that focusing on plastic waste and not the other 99% of waste ensures that we will fail to make any significant progress. It should be obvious that we can't solve a problem by ignoring 99% of it.

PLASTIC HOUSEHOLD WASTE PRODUCTION

Let's look more closely at household waste, even though it is only around 3% of the total. Scientists observed that household waste used to increase every year, then unexpectedly, it stopped increasing. Why were we no longer creating as much waste as expected?

Municipal Solid Waste Management: 1960 – 2018

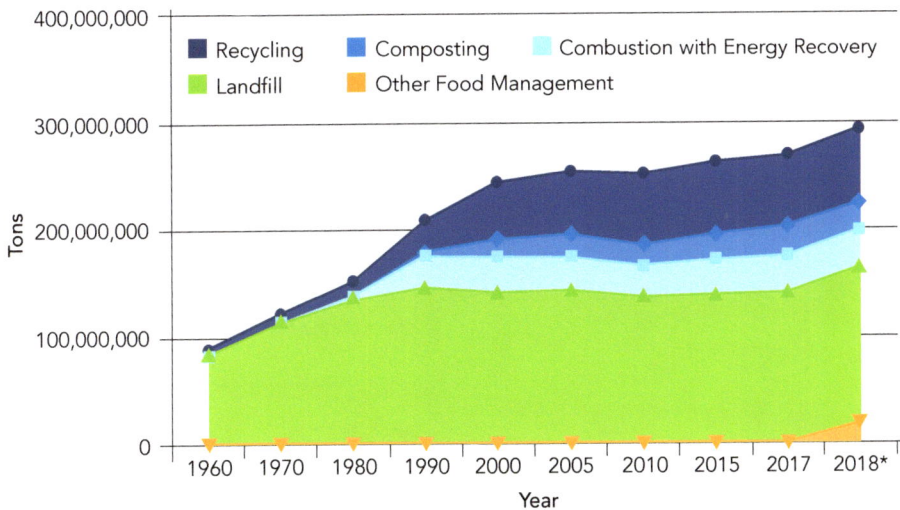

Graph source - https://www.epa.gov/facts-and-figures-about-materials-waste-and-recycling/national-overview-facts-and-figures-materials

It turned out that the growth of plastic use corresponded to a large decrease in the use of other materials.

> *"The increase in plastic waste generation coincides with a decrease in glass and metal found in the MSW stream. In addition, calculating the material substitution rates for glass, metal and other materials with plastics in packaging and containers demonstrates an overall reduction by weight and by volume in MSW generation of approximately 58 % over the same time period."*

D. A. Tsiamis, M. Torres, M. J. Castaldi, Role of plastics in decoupling municipal solid waste and economic growth in the U.S., Waste Management, 77, pp. 147–155, 2018

On average, 1 pound of plastic can replace 3–4 pounds of alternative materials like paper, metal, wood, or glass. It turns out that the net effect of plastics has been to reduce waste creation and waste to landfill. Once more, we see that self-proclaimed "environmental" NGOs have made claims that contradict the evidence.

Another NGO strategy that is in vogue now is the quest for "zero waste." That one baffles me. Why? Try not using the toilet for a week, and you'll see that the "zero waste" quest is a fantasy.

Although waste is natural, normal, and unavoidable, we should still encourage waste reduction and responsible management.

SUMMARY

In this chapter, we have learnt that although we use a lot of plastic by tonnage, it only accounts for less than 1% of materials we use and waste we create (by weight or by volume). Therefore, anyone telling us we need to focus only on plastics while ignoring the other 99% of materials is delusional, ignorant of the facts, or trying to mislead us.

But why would anyone intentionally mislead us about plastics? The answer might surprise you. Competing legacy industries have funded and established NGOs that masquerade as environmental groups, but they actually work to attack plastics, not to protect the environment. This is not speculation; it is a matter of public record. For example, *Beyond Plastics*, a prominent anti-plastic group, is funded by billionaire Michael Bloomberg.

"Bloomberg's new focus on plastics comes at a key moment. The United Nations has projected plastics production will double by 2040, with stark climate implications."

E.A. Crunden, E&E News by Politico, 09/21/2022

Unfortunately, this billionaire forgot to check his facts before acting because, as we have already seen, plastic production increases at the same rate as the other 99% of materials we use. Later in this book, we will see that he was also wrong about the alleged "climate implications." It is a great shame that powerful people take action without doing their due diligence first.

Yes, we should work toward reducing material use and waste across the board. But there's no scientific basis for vilifying plastics. They make up a tiny fraction of the total materials used, plus they reduce overall waste and material consumption when compared to alternatives like paper, wood, metal, or glass.

Now, to examine mismanaged waste, pollution, and litter, and let's see whether the next set of accusations against plastics stands up to closer examination.

MISMANAGED WASTE, POLLUTION & LITTER

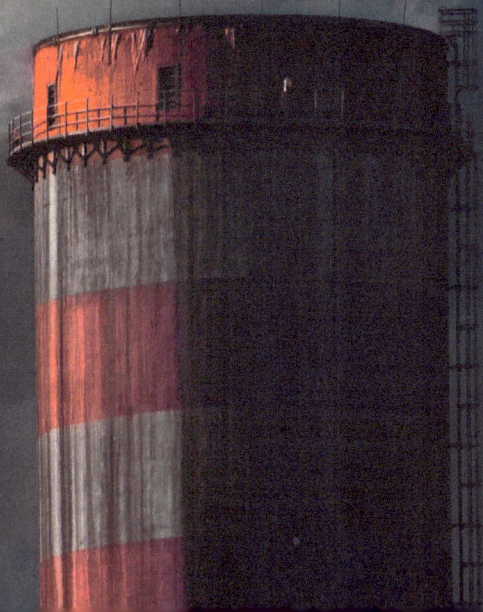

MISMANAGED WASTE, POLLUTION & LITTER

WASTE

Waste — animals make it, and we make it too. It's not new, it's been around for millennia, and it shows no sign of ceasing to exist. We have also managed waste for a very long time. The earliest known wastewater management system was built over 6000 years BC. Burying and burning waste are reliable, low-technology options, and we have developed new methods since then.

It turns out that it's not the amount of waste that we create that matters most, but what we do with it. When it's managed responsibly, all is well. However, mismanagement of waste creates problems. It tends to be unsightly and smelly, and it can even lead to serious health consequences.

We have been told that plastic waste is seriously harming the oceans, but is that assertion supported by scientific studies?

As one might expect, we find that regions with more wealth and larger populations create far more waste.

However, wealthier countries have waste management systems, with bins, collection, recycling, incineration, landfills, and so on. This means that they are not the ones responsible for waste entering the oceans. Why then do the NGOs tell those in the wealthy countries to feel guilty about this issue when it is not actually their waste that causes the most impact?

L. J. J. Meijer et al., More than 1000 rivers account for 80% of global riverine plastic emissions into the ocean, Science Advances, 7, 2021

You can probably guess the answer to that — NGOs tell the wealthier people that they should feel guilty so that they open their wallets and donate to those very same NGOs. The strategy works very well indeed and has made such NGOs hugely wealthy. However, rather than using the money to help, as the donors intended, the NGOs rarely spend any of their money to actually help our environment. Instead, they use it on lobbying and marketing campaigns to attract even more money and increase their influence.

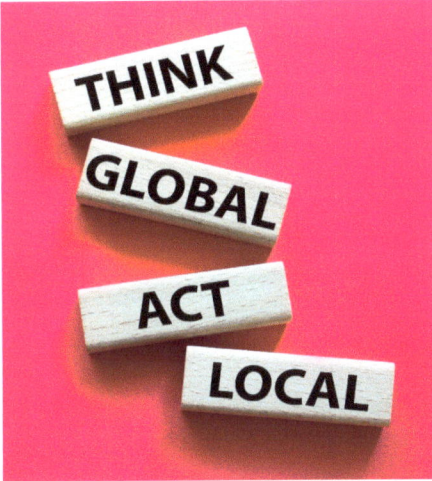

management strategies that can be employed on local scales to reduce inputs."

E. Carpenter & S. Wolverton, Plastic litter in streams: The behavioral archaeology of a pervasive environmental problem, Applied Geography, 84, pp. 93–101, 2017

Carpenter and Wolverton's findings are important because solving the problem of litter requires a different approach than addressing pollution. Solutions for litter that work are education, deposits (to encourage collection), and fines.

People react strongly when they find out that it's not "pollution" caused by companies, but rather litter caused by human litterers. They say, "How can you blame the people? You are just pointing the finger elsewhere to avoid the blame being placed on the plastic industry." — Or words to that effect.

Just as when a parent or a judge decides who is right, the key parameter is evidence. Every party wants to blame someone else, but what does the evidence say? We have plenty of studies on litter, and they show people drop it intentionally. A study where they observed and recorded thousands of littering events came to this conclusion.

"When combined, an estimated 81 % of observed littering occurred with intent."

You read that correctly — over 80 % of littering is intentional, so it is simply wrong to blame the material, the

POLLUTION & LITTER

The term "plastic pollution" is commonplace in the media; even plastic manufacturers use the term. However, scientists have studied that topic too, and they came to a surprising conclusion. They traced the origins of the so-called "pollution" and found out that it was actually "litter." It turns out that litter dropped in one location and later found elsewhere is perceived as "pollution" when it is really just litter that has moved, for example, due to the current in a river or the wind blowing.

"The environmental problem of litter, particularly regarding plastics, is in one sense a local problem that stems from discard behaviors…"

"Correspondingly, we argue that an effective way to reduce the impacts of plastics and other types of litter on aquatic systems is to identify

manufacturer, or the litter itself. Yet, that is what is happening. Situational variables can explain only 15% of the behaviour (such as no bins or existing litter); the rest comes down to the person.

"The results of the current research indicate that 15% of general littering acts result from contextual variables, and 85% result from personal qualities."

P. W. Schultz et al., Littering in Context: Personal and Environmental Predictors of Littering Behavior, Environment and Behavior, 45 (1), pp. 35–59, 2011

Some may say that litter only occurs because there are not enough bins. Studies do indeed show that providing waste receptacles reduces litter, but there is still significant litter, even with waste bins being just 20 feet, that's just 8 steps apart. This is yet more convincing proof that people litter and then look for ways to place the blame elsewhere.

"Further inspection of the data showed that aggregated observed general littering rates were low (and relatively flat at 12%) for receptacles less than 20 feet away. The littering rates increased linearly between 21 and 60 feet and then remained relatively flat at 30% for receptacles 61 feet away and beyond."

Cigarette butts are the most littered item of all. A study found that around 75% were littered and most were not even extinguished, creating a fire risk. That was in an area with an average of 3.5 bins (trash cans) in sight. There is no doubt that this disgraceful behaviour is an intentional, personal choice by the litterers.

V. Patel et al., Cigarette butt littering in city streets: a new methodology for studying and results, Tobacco Control, 22, pp. 59–62, 2013

BLAME

Even with the evidence being crystal clear, there are plenty of allegations of "pollution" where the intent is to blame plastics and companies for the actions of these people we call litterers.

A global study just revealed the world's biggest known plastic polluters

Coca-Cola and PepsiCo came in at the top of a global audit of platic waste

Shannon Osaka, Washington Post, March 24th 2024

And the report from Break Free From Plastic states:

"The analysis reveals that this year's top global plastic polluters are The Coca-Cola Company, Nestlé, Unilever, PepsiCo, Mondelēz International, Mars, Inc., Procter & Gamble, Danone, Altria, and British American Tobacco."

Break Free From Plastics Brand Audit Report 2023
https://brandaudit.breakfreefromplastic.org/brand-audit-2023/

Greenpeace, the Sierra Club, and many others have repeated this outrageous and false claim. Why? Presumably, because it brings them attention and more donations. Does it matter to them that it is untrue? It would appear not.

Here is what the judge had to say about a similar case against Pepsi (New York v. PepsiCo Inc. et al., New York State Supreme Court, Erie County, No. 814682/2023):

"But the judge ruled it would run 'contrary to every norm of established jurisprudence' to punish PepsiCo, because it was people, not the company, who ignored laws prohibiting littering."

Jonathan Stempel, PepsiCo beats New York state's lawsuit over plastics pollution, Reuters, November 1st 2024

The judge also cited precedent that gun manufacturers are not responsible when the gun owner decides to pull the trigger and cause harm. Again, people are responsible, not the company that sold them the product.

Interestingly, I posed this question to my own children's class at the local elementary school. After all, I wrote *The Plastics Paradox* because their teachers had taught my daughters misinformation, so I made sure to go there to teach both the class and the teachers about the evidence. Anyway, I showed the kids a cartoon of a guy who had crashed his car into a tree and asked who was to blame. Should we blame the car, the tree, or the person? Even 8-year-olds got the answer right, so it amazes me that adults struggle to place blame correctly on the person and not the object or the manufacturer. Try buying a Ford car, driving it into a tree on purpose, then arguing to the judge that it was the car's fault or even Ford's. See how convinced the judge is.

DEPOSITS

Science and the legal system agree that littering is caused by people, but there is another, even more powerful way to prove it. There are 8 billion plastic banknotes printed every year — that's one made for every person on the planet every year for decades. However, I have yet to see one on the floor when I go for a walk, or floating in a stream, or deposited on the beach. I often joke that it would be wonderful to go to the beach on holiday and just sit back as the plastic banknotes come rolling in from the ocean. The longer the holiday, the richer I would get!

Why aren't those billions of pieces of plastic littered? Because they have value. That's right — as soon as an object has value, we stop littering it. That's proof that littering is a choice we make, not an accident. It is also the reason deposit systems for cans and bottles work so well. People don't like

dropping items with monetary value. The same applies to credit cards. There are billions of those too; they are made of plastic, and we manage not to drop them.

"We find that a nation-wide DRS can increase PET bottle recycling rates from 24% to 82%, supplying approximately 2700 kt of recycled PET annually. With stability in demand, we estimate that this PET bottle recycling system can achieve 65% bottle-to-bottle circularity, at a net cost of 360 USD/tonne of PET recycled. We also discuss environmental impacts, stakeholder implications, producer responsibility, and complimentary policies toward an efficient and effective recycling system."

R. Basuhi et al., Evaluating strategies to increase PET bottle recycling in the United States, Journal of Industrial Ecology, 28, pp. 916–927, 2024

Not only did the study show that deposits are effective, but it also demonstrated just how effective they are across various geographical locations and deposit sizes.

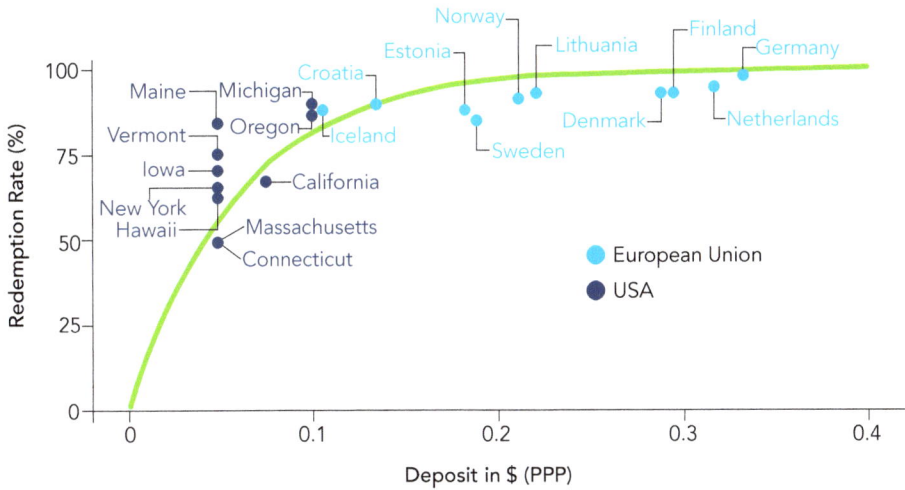

We know definitively that litter is caused by people and that effective solutions include deposits, education, and fines. This study found that fines do work.

"The findings support our hypothesis indicating that littering is more frequent and common when the private cost to littering is not internalised as opposed to when there is a penalty for littering."

F. Salim Khawaja & A. Shah, Determinants of Littering: An Experimental Analysis, The Pakistan Development Review, 52 (2), pp. 157–168, 2013

Blaming materials or companies for litter is unjust, unwise, and counterproductive, but so-called "environmental groups" do it anyway because their greed surpasses any concern they may have once had for the environment. They should be sued, but instead, they are suing innocent companies for litter that was dropped by their customers.

"Legal actions against major corporations escalated in 2023, with lawsuits filed against Danone, Coca-Cola, and Nestlé in Europe. Brand audit data is instrumental in providing evidence for legal battles, underscoring the role of these audits in holding corporations accountable."

https://www.breakfreefromplastic.org/2024/02/07/bffp-movement-unveils-2023-global-brand-audit-results/

MISCONCEPTIONS

Before we move on, it seems like the right time to clear up some other common misconceptions about litter. On social media, people often demand more recycling of plastic as a solution to litter, aka "pollution," but there is no link between the two. People drop litter intentionally, as we have seen, and there is no evidence that they change their behaviour because of local recycling rates.

Another common belief is that plastic litter is only a problem because the plastic is not degrading quickly enough. People literally drop litter and then have the audacity to blame that litter for not vanishing, like they expect a wizard to wave a wand and clean up their mess for them. This misguided thought is responsible for all the buzz around degradable plastics and other degradable materials.

However, there are serious errors with this line of thinking. Firstly, a paper bag weighs 60 g compared to a plastic bag that weighs just 6 g. So, banning plastic bags and changing to paper means a tenfold increase in the weight of litter. This is an extreme case, but as we saw earlier, replacing plastic means 3–4 times more material, and therefore, a much greater mass of litter.

Secondly, degradation means the conversion of solids into carbon dioxide, which is exactly what most people are against.

Plus, people want degradable materials because they believe that normal polyethylene bags do not degrade. NGOs, including the World Wildlife Fund, tell us that it takes hundreds of years for a plastic bag to degrade. However, they tell us that without a shred of evidence, and it turns out to be just another internet myth spread by NGOs for their own nefarious purposes. As we shall see in a later chap-

ter, it is factually incorrect to say that plastics do not degrade. In fact, they do degrade and far more rapidly than people realise.

Thirdly, studies show that when you replace plastic with a material that the public believes to be degradable, they drop much more litter. A study compared PET bottles to paper-based drink cartons.

"As was expected, the PET bottles were littered less than the Cartocans. Of the PET bottles, 2.6% was littered, while 5.8% of the Cartocans was littered…"

"In addition to this, 16 of the peel-off closures of the Cartocan were found littered, while none of the caps of the PET bottle were found separately."

R. Wever et al., Influence of Packaging Design on Littering and Waste Behaviour, Packaging Technology and Science, 23, pp. 239–252, 2010

Recently, companies have been touting their new tethered caps, which are attached to the PET bottle to prevent littering in the same way that "idiot mittens" work (mittens tied together with a string through the arms of the jacket to prevent us from losing them). It appears that such tethering may be of limited utility.

BEACHES

For some reason, people are especially interested in beach litter. Perhaps because beaches are so beautiful and litter is so jarring against a pretty background. Whatever the reason, scientists have also studied beach litter.

For popular beaches, they found that litter does not come from the oceans, as some believe, but mainly from the people on the beach. Apparently, people litter until the beach is ugly, then go find a new clean beach to ruin. That is human behaviour, and the solution lies in altering that bad behaviour through education, deposits, and fines.

> *"Beach users have been shown to be main contributors to debris along coastal and estuarine shores. The ease of access to a beach is a main factor to the number of beach visitors."*

K. Willis et al., Differentiating littering, urban runoff and marine transport as sources of marine debris in coastal & estuarine environments, Nature Scientific Reports, 7, 44479, 2017

> *"The beaches with lower levels of urbanization also had smaller quantities of anthropogenic litter. Items related to beach users were predominant for most of the beaches. The confirmation that beach users are primarily responsible for the generation of anthropogenic litter may contribute to the development of strategies to reduce the problem, such as installing bins and distribution containers for anthropogenic litter collection and designing educational campaigns for beach users."*

M. C. B. Araújo et al., Anthropogenic Litter on Beaches With Different Levels of Development and Use: A Snapshot of a Coast in Pernambuco (Brazil), Frontiers in Marine Science, 5 (233), 2008

The study in Brazil found many wooden sticks on the beach. I visited São Paulo to give a keynote, and my friend Evandro explained that those sticks are used to eat cheese. This uniquely local type of litter helps scientists to prove that it was caused by the people on the beach and not washed up from elsewhere.

One recurring response from people when talking about litter is an offer for me to fly to Malaysia, Indonesia, Hawaii, or the Philippines and see that I'm wrong. That response is especially illogical and counterproductive. Firstly, no one claimed that litter doesn't exist. In places where people drop a lot of litter, we find a lot of litter. The people there cause it, and the solution is behavioural change. The second reason that the suggestion is crazy is that the impact of flying to see litter has the same greenhouse gas impact as 10,000–20,000 PET bottles, which is more bottles than I will use in my entire lifetime.

People seem to think that flying around or sailing to see environmental degradation is noble and justifiable, but it is neither. Certain rich CEOs do it, and I'm sure it makes them feel and look good, but there is no scientific reason to do it because we already have the studies and decades of data available on our laptops with no travel needed. It's ironic that people's reaction is to do what's worse for the environment, which is flying around to look at it.

What about beaches without people on them? Remote beaches also face some contamination, but the amount of litter is far lower and is mainly composed of improperly discarded fishing gear, including nets washed up on the tide.

Some remote beaches still have large quantities of consumer items. How can that be when there are no people there to drop litter? Scientists have investigated that too, and by examining the litter in detail then doing some detective work, they discovered the culprits are fishing vessels throwing trash overboard.

"Many oceanic islands suffer high levels of stranded debris, particularly those near subtropical gyres where floating debris accumulates. During the last 3 decades, plastic drink bottles have shown the fastest growth rate of all debris types on remote Inaccessible Island. During the 1980s, most bottles drifted to the island from South America, carried 3,000 km by the west wind drift. Currently, 75% of bottles are from Asia, with most from China. The recent manufacture dates indicate that few bottles could have drifted from Asia, and presumably are dumped from ships, in contravention of International Convention for the Prevention of Pollution from Ships regulations. Our results question the widely held assumption that most plastic debris at sea comes from land-based sources."

P. G. Ryan et al., Rapid increase in Asian bottles in the South Atlantic Ocean indicates major debris inputs from ships, Environmental Sciences, 116 (42), pp. 20892–20897, 2019

SUMMARY

In the previous chapter, we saw that plastic is less than 1% of both materials waste and total waste by weight or volume. So, for waste, a focus on plastics cannot deliver meaningful results. Scientists have argued that focusing on less than 1% of waste distracts from addressing more than 99% of waste, where we could make a real difference. Replacing plastics with alternatives creates 3–4 times more waste, so that would not be a positive move. Instead, we should make choices that reduce impact based on the data.

In this chapter, we have seen that although countries with greater wealth and higher populations generate more waste, that waste is, for the most part, properly managed. Other countries have not yet caught up, and those that dump their waste on land and into rivers are generating a disproportionately large and negative impact, especially for our oceans. The solutions are known and working in many countries, so we need to help them catch up. Sometimes, I hear the argument that those countries do not have the technology to manage their waste. However, we have been managing waste for millennia, e.g. by burying or burning it — technology is not the obstacle.

Scientists have determined that the term plastic "pollution" is inaccurate; what people consider pollution is really "litter" that has moved and accumulated in other places, such as rivers or oceans. So, while NGOs and even plastic manufacturers mistakenly talk of "pollution," a problem associated with industry, the actual culprit is litter. This revelation has important consequences because the correct solution depends on accurately diagnosing the problem. The discovery that litter is the issue allows us to implement proven solutions, such as deposits, education, and fines. Singapore is super clean because of severe fines, whereas Japan is clean due to their culture, which takes a strong stance against litter.

Once more, we have seen how NGOs have misled us by blaming plastics and companies when science and the courts agree that litter is caused by people. Now, we need to make sure that the public and our policymakers recognise the truth so that appropriate policy can follow.

We know bans will not solve the litter crisis. There is no evidence that a litterer will choose to drop a PET bottle but not drop a metal can. We know such people will misbehave no matter what material is used, and alternative materials lead to increased amounts of litter, not less.

Extended producer responsibility (EPR) has been suggested as a solution, but we know that is not the answer either. Firstly, because it is not the producer who drops the litter, nor is it their responsibility to pick it up. Secondly, we have already paid taxes for litter cans, cleanup, and disposal. EPR systems would force us to pay twice for the same service — once in our taxes and a second time in increased prices because the manufacturers will certainly pass the cost of EPR to consumers.

Lastly, let us remember why people litter plastic more than some other materials. It is because it is so inexpensive. The lower the cost, the more tempting it is to litter. The answer is not to move to alternatives that both cost more and increase impact, but rather we should stay with our cheapest, greenest option and encourage people to act responsibly.

The current prevailing attitude is just the opposite. We are being told to limit production and access to our cheapest, greenest option. That may be fine for wealthy people, but what does it mean for the poor? They cannot afford more expensive options. They buy food by the sachet because one sachet is all they can afford. Wealthy people telling others what to do is unjust and counterproductive. People should be free to make personal choices but not to inflict their ideologies on others.

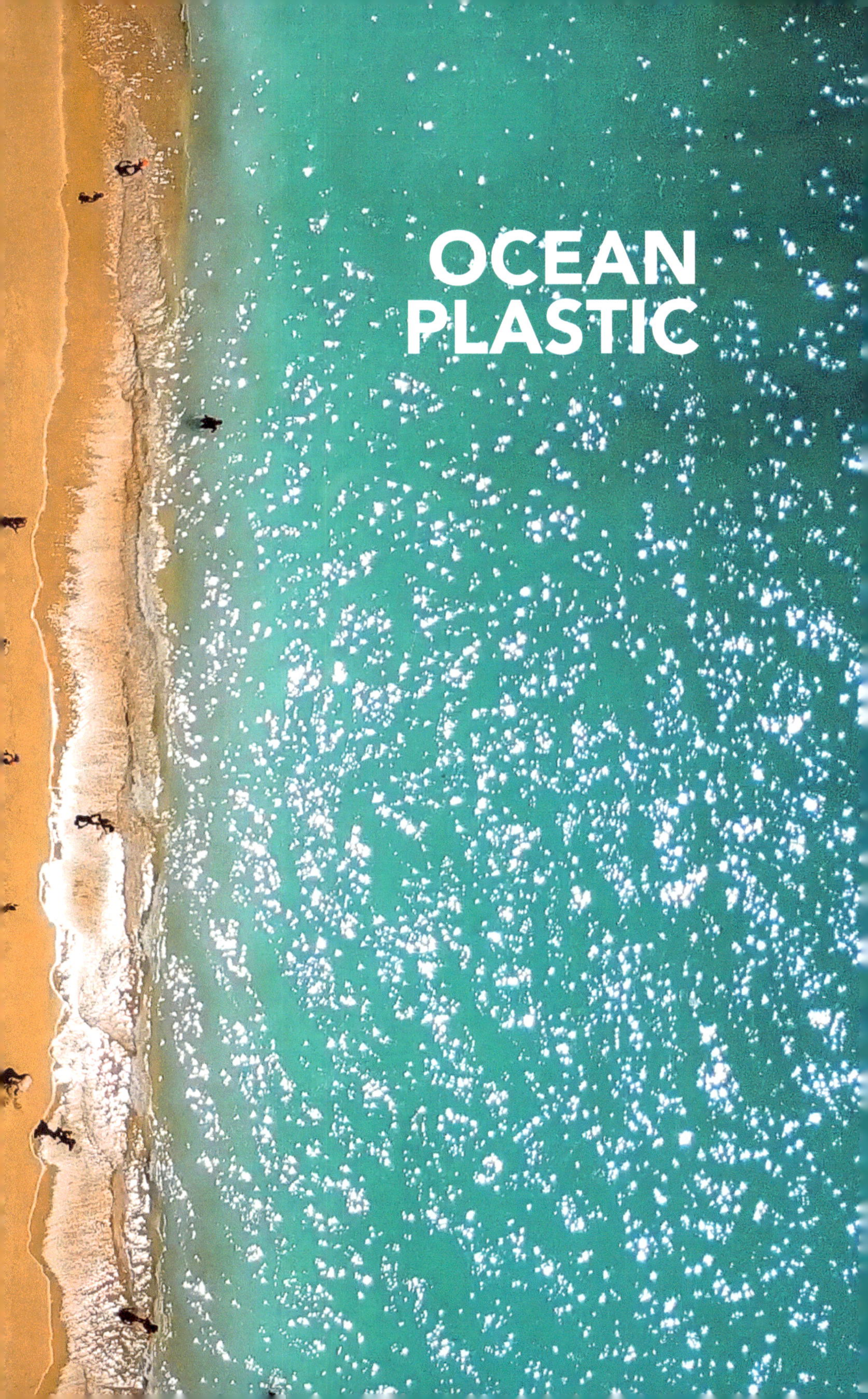
OCEAN
PLASTIC

OCEAN PLASTIC

So much has been said about plastics and their effects on our oceans. Allegedly, there is a floating island of plastic the size of Texas. Not only that, but, apparently, over 10 million tons of plastic enter oceans every year, "choking" them and causing harm to marine animals. It has even been claimed that there will be more plastic than fish in the ocean by 2050. People are demanding action, and I don't blame them. However, as a scientist, I know it is wise to check the facts before jumping into action. Fortunately, there are many studies on this topic, so we have the information we need to know what is going on and what to do about it. Let us examine each claim one by one, comparing them to what scientists have to say.

FLOATING ISLAND OF PLASTIC

Here is a quote from a PhD thesis that investigated the floating island of plastic myth and how it came to be.

"Time Magazine describes a 'swirling mass of plastic debris twice the size of Texas,' human impact on the ocean so severe 'You can literally see the result' (Walsh 2008). The garbage patch is crowned 'The World's Largest Landfill' by Discover amidst calls to recognize it as 'the 8th continent' (Kostigen 2008). Visible. Solid. Massive. The collective account does not shy from specifics. As reported by ABC News, the San Francisco Chronicle, and even Oprah, among countless others, the garbage patch spans hundreds of miles, is one hundred meters deep, and weighs 3.5 million tons (Berton 2007; Bonfils 2008). It is, following the most recited descriptor, twice the size of Texas. Or, in all its regional variations, 'as large as Central Europe' (Pravda 2004), with a 'footprint as large as France and Spain combined' (WHIM 2014), even 'twice the size of America' (Daily Mail 2008). This floating mass growing in the North Pacific Ocean, northeast of the Hawaiian Islands is surely impossible to miss...'"

And now for the punchline...

"But despite general agreement on its location and the proliferation of claims about its size, no one can find it; not on Google Earth, not after weeks at sea. The trash island is not there."

K. De Wolff, Gyre Plastic: Science, Circulation and the Matter of the Great Pacific Garbage Patch, PhD Thesis, University of California, San Diego, 2014

The researcher interviewed Captain Moore who discovered the "patch" to ask him how the myth of an island was created, and this is what she discovered.

"My next question, and what I so desperately want to know, is who was the first to call the garbage patch a trash island? To my surprise, Moore points to 'foreign papers,' specifically Pravda and proceeds to describe a captivating image of a 'Matterhorn looking mountain,' an artists conception of a floating trash heap. Soon after the interview, I search media archives and am pleased to find that Moore's tip checks out — the earliest mention of a floating trash island does appear to be in Pravda Online, February 24th, 2004. The short article, '"Trash Island" discovered in the Pacific Ocean,' takes its content in turn from an article in German National Geographic equivalent Geo that describes a 'carpet' of plastic in the ocean. How the carpet turned into an island remains a mystery of English-German-Russian-English translation."

So, the German *National Geographic* reported a "carpet" of plastic, but the Russian newspaper Pravda translated "carpet" as "island" accompanied by an artist's impression of a mountain of plastic. This is how easy it is to create a myth that rapidly spreads around the world with no one thinking to check whether it's actually true. Ironically, "pravda" means "truth" in Russian.

In his book *Plastic Ocean*, Captain Moore described what the gyre is, in his own words:

"Let it be said straight up that what we came upon was not a mountain of trash, an island of trash, a raft of trash or a swirling vortex of trash — all media-concocted embellishments of the truth. It would become known as the Great Pacific Garbage Patch a term that's had great utility but, again,

suggests something other than what's out there. It was and is a thin plastic soup, a soup lightly seasoned with plastic flakes, bulked out here and there with 'dumplings': buoys, net clumps, floats, crates, and other 'macro debris'."

Plastic Ocean: How a Sea Captain's Chance Discovery Launched a Determined Quest to Save the Oceans, C. Moore & C. Phillips, Avery/Penguin 2011

If there is no floating island, then what about the "soup"? How much plastic is there? The number we see everywhere and touted by NGOs is 10—12 million tons of plastic entering the oceans per year. They often express it as a truckload of plastic per minute. Here's one headline from Greenpeace, and CNN picked up the story, along with many more.

Every minute of every day, the equivalent of one truckload of plastic enters the sea

Jen Fela

https://www.greenpeace.org/international/story/15882/every-minute-of-every-day-the-equivalent-of-one-truckload-of-plastic-enters-the-sea/

That sounds like a lot, but where does this number come from and how accurate is it? The estimate comes from an old publication by Jambeck. It had an enormous impact and has been quoted widely ever since.

"Plastic debris in the marine environment is widely documented, but the quantity of plastic entering the ocean from waste generated on land is unknown. By linking worldwide data on solid waste, population density, and economic status, we estimated the mass of land-based plastic waste entering the ocean. We calculate that 275 million metric tons (MT) of plastic waste was generated in 192 coastal countries in 2010, with 4.8 to 12.7 million MT entering the ocean."

J. Jambeck et al., Plastic waste inputs from land into the ocean, Science, 347 (6223), pp. 768–771, 2015

But there's a major problem with the publication: It's pure guesswork, totally unsupported by any kind of data! How was the estimate made then? The author estimated the amount of unmanaged waste and then assumed that a large proportion of it gets into the rivers and is washed into the ocean. The authors admit that accurately estimating the actual amount is impossible; therefore, they guessed that up to 45% of mismanaged waste somehow reaches the ocean.

"Some percentage of the total mismanaged plastic waste (inadequately managed plus litter) enters the ocean and becomes marine debris. To our knowledge no direct estimates of this conversion rate exist."

J. Jambeck et al., Plastic waste inputs from land into the ocean, Supplemental Material, Science, 347 (6223), 2015

Jambeck made her guess in 2015, and in the years since, scientists have scoured the oceans looking for the millions of tons of plastic that she says should be there. But they failed to find it. "Where is the 'missing plastic'?" they asked. The most comprehensive analysis comes from Weiss, who meticulously analysed all the data on ocean plastic collected over the years by various research groups. They pointed out that Jambeck claimed rivers are the major source of ocean plastic.

"Leakage from waste generation and inadequate disposal on land—i.e., mismanaged plastic waste (MPW)—was initially identified as the main driver for plastic discharge to the ocean, with a potential annual transfer of 4.8 to 12.7 million metric tons (Mt). Rivers are recognized as the principal conveyors in this transfer."

And Weiss goes on to say that the actual amounts of plastic coming from the rivers are a thousand times less than Jambeck claimed: not 10 or 12 million tons, but actually about 6000 tons a year.

"On the basis of an in-depth statistical reanalysis of updated data on microplastics—a size fraction for which both ocean and river sampling rely on equal techniques—we demonstrate that current river flux assessments are overestimated by two to three orders of magnitude."

L. Weiss et al., The missing ocean plastic sink: Gone with the rivers, Science, 373 (6550), pp. 107–111, 2021

It is vital to stress that this is not one study against another study. This is one study based on a guess against many other independent studies spanning many years and many thousands of actual measurements.

Even back in the same year that Jambeck came up with the millions-of-tons guess, other scientists pointed out that it didn't agree with the evidence. Cózar showed that ocean plastic is hundreds or thousands of times less than Jambeck stated and that the 10-million-ton guess is wildly too high.

"In the present study, we confirm the gathering of floating plastic debris, mainly microplastics, in all subtropical gyres. The current plastic load in surface waters of the open ocean was estimated in the order of tens of thousands of tons"

"Nevertheless, even our high estimate of plastic load, based on the 90th percentile of the regional concentrations, is considerably lower than expected, by orders of magnitude."

A. Cózar et al., Plastic debris in the open ocean, PNAS, 111 (28), pp. 10239–44, 2014

Scientists even went so far as to say that people with ulterior motives have intentionally misled us about ocean plastics.

"In this viewpoint, we argue that plastic pollution has been overemphasised by the media, governments and ultimately the public as the major threat to marine environments at the expense of climate change and biodiversity loss. We discuss why this can be a convenient truth, especially as some mechanisms to reduce plastic waste play into corporate greenwashing in a neoliberal economy rather than addressing the root cause of overconsumption of resources."

R. Stafford & P. J. S. Jones, Viewpoint - Ocean plastic pollution: A convenient but distracting truth?, Marine Policy, 103, pp. 187–191, 2019

They state that the focus on plastics is an attempt to distract us from the real problems, such as the overconsumption of resources.

MORE PLASTIC THAN FISH BY 2050?

The Ellen MacArthur Foundation made the claim that there would be more plastic than fish in the ocean by 2050.

"The best research currently available estimates that there are over 150 million tonnes of plastics in the ocean today. In a business-as-usual scenario, the ocean is expected to contain 1 tonne of plastic for every 3 tonnes of fish by 2025, and by 2050, more plastics than fish (by weight)."

Other organisations, including the WWF, WEF, Greenpeace, Plastic Soup, Surfers Against Sewage and UNEP, have repeated the claim. It's a scary

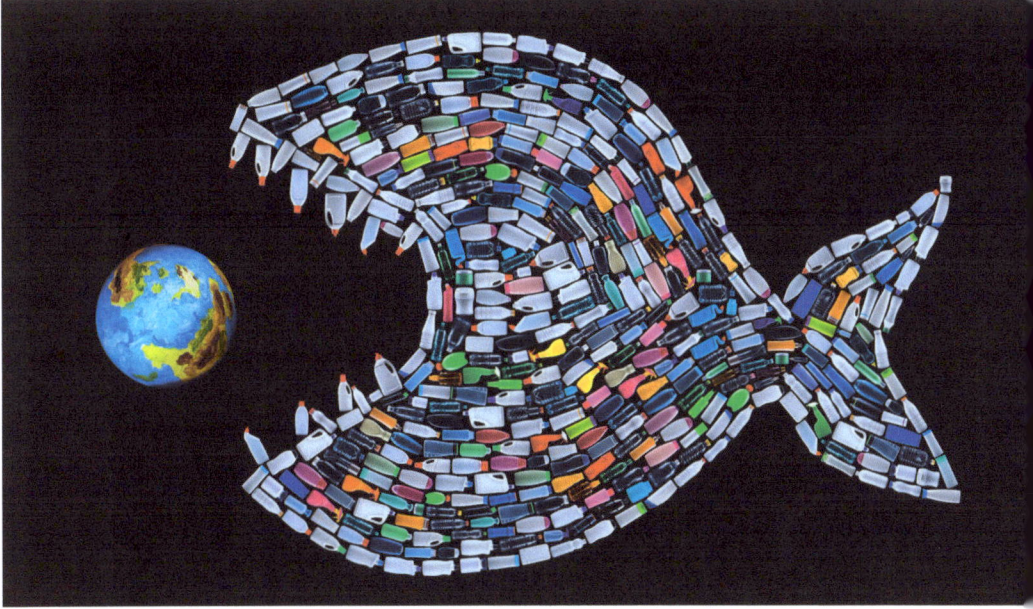

thought that captures our imagination and sticks in our minds. But is it true?

Here are the assumptions they made.

- First, they claim that there are 150 million tons of plastic in the oceans already.
- Secondly, they claim that the amount of plastic grows by over 8 million tons per year.
- Thirdly, they claim the amount is growing exponentially.
- Fourthly, implicit in their calculation is that none of the plastic degrades and vanishes.
- Finally, they claim that the total amount of fish in the oceans is 800–900 million tons.

The problem is that every assumption they made is wrong. In fact, the BBC and the CBC both showed that the claims were shaky at best, and a closer scientific examination totally discredits the claim as pure fiction. There is no evidence that there are 150 million tons in the ocean now. They used the disproven Jambeck estimate for the amount added per year. Then, they assumed, without proof, that the amount increases exponentially despite 50 years of data across many studies in this comprehensive review showing no such increase in the amounts of plastic in the ocean, on beaches, or on the ocean floor.

"For microplastics, floating particles were found at similar levels between 2005 and 2014 in East Greenland,

in the North Atlantic Subtropical Gyre between 1986 and 2008 and in the North Pacific Subtropical Gyre between 2001 and 2012. In addition, no changes in floating microplastics (>150µm) were detected between 1987 and 2015 in the Baltic Sea, between 1987 and 2012 in the North Atlantic subtropical gyre and between 2001 and 2012 in the North Pacific Subtropical gyre. For ingested large debris, constant levels were also demonstrated for stranded cetaceans recorded from Irish waters between 1990 and 2015, and in western Mediterranean sea turtles between 1995 and 2016."

F. Galgani et al., Are litter, plastic and microplastic quantities increasing in the ocean?,
Microplastics and Nanoplastics, 1 (2), 2021

They also found no increase in microplastic or ingested plastic since 2000.

"For large debris on beaches, an absence of temporal trend was demonstrated for macroplastics in the North Atlantic, between 2001 and 2011, in Chile, between 2006 and 2016 and for data from cleanups in Taiwan, between 2004 and 2016. An absence of temporal trends was also observed for large floating debris in the Balearic Islands between 2005 and 2015 and in China, between 2007 and 2014. In addition, collections of marine litter by Continuous Plankton Recorders showed relatively unchanged amounts trapped annually in the North East Atlantic since 2000, following a steady increase since the 1950s."

"In seafloor litter studies, no change in plastic pollution was measured in Spain between 2007 and 2017 nor in the North Sea. A slight increase in seafloor plastics was observed in recent years in the Baltic (excluding fishing gear), while results from observations in France, between 1995 and 2017 (23 years), showed mixed trends, of decreasing amounts between 2000 and 2013 and of increases since 2013. No trend was identified in Chinese waters for sea floor litter between 2007 and 2014, with a large variability in plastics concentration and from data collected during regular State monitoring between 2011 and 2018. In contrast, a decrease in total seafloor litter was measured between 2007 and 2017, in both the Alboran Sea and the northern Adriatic, without significant temporal trends for plastic in the remaining Adriatic."

The review is incredibly thorough, covering study after study across decades, yet the data consistently shows that the amount of ocean plastic is not increasing. This stands in stark contrast to the narrative being pushed by policymakers and NGOs. These groups often rely on modelling studies that predict an increase, even when overwhelming real-world data shows that the models are flawed.

Why is this discrepancy ignored? Corrupt NGOs seem unwilling to let inconvenient facts disrupt their agenda. A genuine environmental organisation would celebrate and share this positive news, yet we've seen no such behaviour — have we?

What about the notion that the plastic just accumulates and never degrades? Is that true?

"I was shocked by how small the pieces were. I was shocked that so many pieces were so tiny and that everything was degrading so quickly."

K. De Wolff, Circulating Away: Plastic, Science and the Great Pacific Garbage Patch, PhD Thesis, University of California, San Diego, 2014

Subsequent chapters cover degradation and microplastics, demonstrating that plastics do degrade and do so much more rapidly than anyone imagined. We will not go into more detail on those subjects here except to point out what scientists have said about the effects of microplastics on the ocean.

"We conducted an ecological risk assessment of MP [microplastics] in the global ocean by comparing the thresholds of biological effects with the probability of exposure to those concentrations…"

"Levels of MP from 100 to 5000 μm span from < 0.0001 to 1.89 mg/L, whereas the most conservative safe concentration is 13.8 mg/L, and

probability of exposure is p = 0.00004. Therefore large MP pose negligible global risk."

R. Beiras & M. Schönemann, Currently monitored microplastics pose negligible ecological risk to the global ocean, Nature Scientific Reports, 10, 22281, 2020

So, there is "negligible" risk because there are simply far too few microplastics in the ocean to have any effect, and as we saw previously, the amount is not increasing.

Here is another study that shows no accumulation of microplastics and 10,000 times too few microplastics to cause any effect.

"Microplastics are ingested and, mostly, excreted rapidly by numerous aquatic organisms. So far, there is no clear evidence of bioaccumulation or biomagnification."

"Based on the evaluated data, the lowest concentrations eliciting adverse effects in aquatic organisms exposed via the water are by a factor of approximately 10 000 times than maximum microplastic concentrations found in marine waters."

K. Duis & A. Coors, Microplastics in the aquatic and terrestrial environment: sources (with a specific focus on personal care products), fate and effects, Environmental Sciences Europe, 28 (2), 2016

The fact that the debunked claim that the oceans contain more plastic than fish (at least by 2050) has never been retracted is very telling. Organisations genuinely dedicated to helping the environment would publish accurate information and retract any claims that turned out to be false because we can only make wise decisions based on accurate data. When self-proclaimed environmental groups perpetuate falsehoods, it makes one question their credibility and what their real motives are.

OCEAN CLEAN-UP

Since most people imagine a floating island of plastic, they think you can just go there and scoop it up or tow it away. Having read this book, you realise that there is no floating island and that the small pieces of plastic, which are widely dispersed, mean that clean-up is not realistic.

"A lot of people hear the word patch and they immediately think of almost like a blanket of trash that can easily be scooped up, but actually these areas are always moving and changing with the currents, and it's mostly these tiny plastics that you can't immediately see with the naked eye." — Diana Parker, National Oceanic and Atmospheric Administration

oceanservice.noaa.gov/podcast/mar18/nop14-ocean-garbage-patches.html

In fact, the National Oceanic and Atmospheric Administration (NOAA) has made some calculations on the idea of cleaning up the gyres with ships and nets.

"We did some quick calculations that if you tried to clean up less than one percent of the North Pacific Ocean it would take 67 ships one year to clean up that portion. And the bottom line is that until we prevent debris from entering the ocean at the source, it's just going to keep congregating in these areas. We could go out and clean it all up and then still have the same problem on our hands as long as there's debris entering the ocean."

How much would it cost to attempt a clean-up using ships? Here are some calculations I found online.

"Suppose we were to attempt to clean up less than 1% of the North Pacific Ocean (a 3-degree swath between 30° and 35°N and 150° to 180°W), which would be approximately 1,000,000 km². Assume we hired a boat with an 18 ft (5.5 m) beam and surveyed the area within 100 m off of each side of the ship. If the ship traveled at 11 knots (20 km/hour), and surveyed during daylight hours (approximately 10 hours a day), it would take 67 ships one year to cover that area! At a cost of $5,000-20,000/day, it would cost between $122M and $489M for the year. That's a lot of money—and that's only for boat time. It doesn't include equipment or labor costs (keep in mind that not all debris items can be scooped up with a net)."

Carey Morishige, Pacific Islands Regional Coordinator, NOAA Marine Debris Program

Not only would this approach fail, but it would also come at an enormous financial cost. And let us not forget the diesel fuel burned and the black smoke emitted by all those ships. The net effect (pun intended) would be to increase environmental harm. The "cure" is worse than the illness, but that has not stopped organisations from raising huge sums to do just that. One has to wonder if they are fraudsters who are fully aware of the facts and don't care about profiting off good-hearted, but gullible, donors.

Gyre Plastic

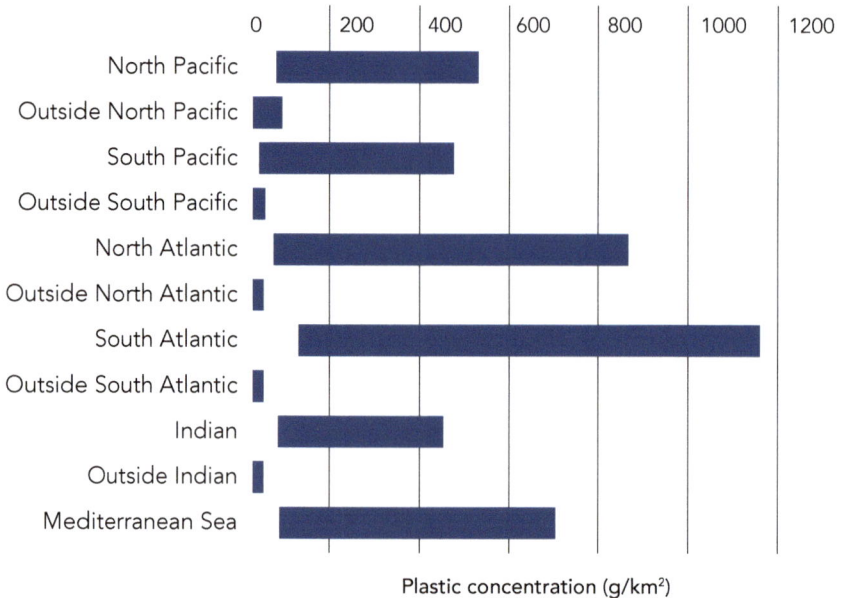

Plastic concentration (g/km²)

A. Cózar et al., Plastic Accumulation in the Mediterranean Sea, PLOS ONE, 10 (4), 2015

There is so much talk about the gyres that no one ever looks at the amount of plastic and other debris outside them in the rest of the ocean. Scientists have studied that too, though, and the answer is that the amount of debris is close to zero. Yet, good news does not make headlines or get donations into the greedy hands of NGOs, which is why this information has not been shared more widely. Even within those areas where ocean currents concentrate debris, the maximum amount of plastic is about 1 kg (about 2 lbs) per square kilometre. That means that the oceans are not "choking in plastic."

Another term used for the gyres is a "plastic soup," but again, that is not accurate. If it is a soup, then it is the wateriest, most useless soup you have ever seen. The amount of plastic would be one tiny flake per three bathtubs of clean ocean water.

Don't misunderstand my point: There should not be any chemicals, metal, paper, glass, or plastic in the ocean, and we should stop dumping these things there where they do not belong. At the same time, it is counterproductive to lie to the public, our children, and policymakers about the amounts. We need to stop the lies and start making rational decisions based on real data.

WHAT IS IN THE GYRES?

There is so much talk about the gyre. We now know that there is no floating island of plastic and that the gyres are actually areas where the ocean currents concentrate floating materials. So, what exactly is there? Is there actual harm caused? And what should we do, if anything?

"Our new results indicate that a significant fraction of these hard plastics may also be coming from fishing vessels. Adding to the mass of floating nets and ropes, this suggests that between 75 and 86% of the floating plastic mass (> 5 cm) in the NPGP could be considered abandoned, lost or otherwise discarded fishing gear."

L. Lebreton et al., Industrialised fishing nations largely contribute to floating plastic pollution in the North Pacific subtropical gyre, Nature Scientific Reports, 12, 12666, 2022

They recorded over 500 kg of material, totalling 6048 items. The vast majority were discarded fishing gear, including so-called "ghost nets" that harm marine life like whales, turtles, and fish. How many bags did they find? Zero. How many straws? One. How many plastic bottles? Nine. Those common consumer items made up only 0.03% of the material found!

The United Nations have been holding many meetings to address this problem. That sounds wonderful until you read their plans, which are to completely ignore the fishing nets that cause harm and instead focus on the 0.03% of consumer items that do not. That is a sure path to failure and makes one wonder about the competence and true motives of the United Nations Environment Programme (UNEP).

Also, it's worth mentioning that the study noted the total amount had decreased by 33% since the last measurement 4 years prior. Of course, no one was told about that because good news doesn't warrant a mention and it doesn't create donations for NGOs.

Now we know what is there, it is time to look at what really harms marine life.

TURTLES

What about the famous video claiming that a turtle had a straw up its nose? That has been watched about 200 million times on YouTube and helped to propel an anti-plastic straw movement resulting in bans in many countries. The problem with that is that there is no evidence that the item was made of plastic or that it was a straw. They pulled out an object and asked each other what it was.

Man: *"Do you know what this is?"*
Woman: *"What is it?"*
Man: *"That's a worm."*
Woman: *"Is it a hookworm?"*
Man: *"I think it's a tube worm."*

Eventually, they pull the object out and say, "He bit on it, and he said it's plastic." That is all the "proof" ever given that it was plastic — some dude in swimming trunks bit on it. As a scientist, I can reveal that is not a valid test. To know what something is made of, we need to send it to a laboratory for tests like infra-red spectroscopy (FTIR), Raman spectroscopy, or nuclear magnetic resonance (NMR). Biting does not count. When you look at the object, it is not at all clear what it is. So, a fellow scientist wrote and asked the sea turtle video "research-ers" how they knew it was a straw. They replied:

"Thank you kindly for reaching out. I can confirm that we did not run any chemical tests to 100 % confirm the nature of the 'straw'. I doubt that it is surgical PVC or anything similar, although I think there is a slim possibility that it could be electrical wire insulation."

Nathan J. Robinson, Marine Biologist and Science Communicator, 19 July 2021

200 million people believe a sea turtle was harmed by a straw despite there being zero evidence that it was a straw and zero evidence that it was made of plastic. Welcome to a world where crazy nonsense travels all over the world, evoking emotions and actions that make things worse. As we will see later in the book, moving to paper straws increases impact and does nothing to help the oceans because it was never an issue in the first place.

What about bags? Are they a threat to turtles? After all, we see images of turtles with bags around their necks or eating plastic bags almost daily. It turns out that every such image we have ever seen was made in Photoshop or similar image editing software.

The BBC showed such an image, and I busted them on social media for it. To their credit, they apologised and updated the story as shown.

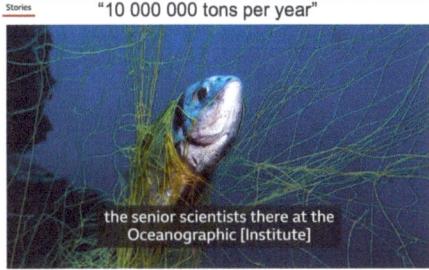

"Correction 16th January 2023: The article originally included a picture of a Hawksbill Turtle swimming underwater while entangled in a plastic bag. However as this was a concept picture, and the bag was not present in the original photo, we have replaced this image."

https://www.bbc.com/news/av/stories-64250382, 14 January 2023

Surprisingly, there is an ongoing experiment that reveals the incidence of plastic bags in the ocean. They have been sailing ships all around to measure plankton and every time their small net gets clogged with a piece of plastic bag, they have to remove it and make a note in the ship's logbook. The remarkable part is that this same method had been used for decades from 1957 to 2016, spanning 6.5 million nautical miles of ocean travelled. How far does the ship have to sail before it collects a piece of plastic bag in the net? Have a guess…

The answer is 100,000 miles. That is 4 times around the planet just to find one piece of bag, so anyone telling us that the ocean is drowning in plastic bags is not fond of honesty. Not only that, but they reported peak amounts a few decades ago followed by a decrease.

C. Ostle et al., The rise in ocean plastics evidenced from a 60-year time series, Nature Communications, 10 (1622), 2019

This is not the only such measurement either. A recent collection of items for analysis of the Pacific Ocean Gyre found 0.00 % bags. Junk is in the oceans, and it should not be there, but let us take actions that matter rather than making up fiction about it.

"The composition of marine debris...was similar to that found in other studies for the western Mediterranean and their amounts seem not to be an important threat to turtle survival in the region."

Source of Mortality Caused by Humans	Mortalities per Year
Shrimp trawling	5,000 - 50,000
Fishery (trawl & release, passive gear, net entanglement)	500 - 5000
Collisions with boats	50 - 500
Dredging	5 - 50
Other	20 - 200

National Research Council - Decline of the Sea Turtles: Causes and Prevention. Washington, DC: The National Academies Press, 1990

Allen M. Foley et al., Characterizing Watercraft-Related Mortality of Sea Turtles in Florida, The Journal of Wildlife Management, 83 (5), pp. 1057–1072, 2019

F. Domènech et al., Two decades of monitoring in marine debris ingestion in loggerhead sea turtle, Caretta caretta, from the western Mediterranean, Environmental Pollution, 244, pp. 367–378, 2019

Studies on what harms turtles exist, and they show, as expected, that it is abandoned nets. The same nets that UNEP plans to ignore.

WHALES

What is a threat to whales? Again, we have multiple studies and not one mention of the words: "plastic," "bag," "bottle," or "straw." Anyone truly out to protect whales should push for regulations on abandoned fishing nets and better sonar for boats to prevent them from running over the whales. Logical, helpful solutions come into focus once we have the data.

Mortality All Causes	Mortalities per Year
Entanglement in fishing gear	323
Natural causes	248
Vessel strikes	171

J. M. Van der Hoop et al., Assessment of Management to Mitigate Anthropogenic Effects on Large Whales, Conservation Biology, 27 (1), pp. 121–133, 2012

R. Knowlton, S. M. Kraus, Mortality and serious injury of northern right whales (Eubalaena glacialis) in the western North Atlantic Ocean, , Journal of Cetacean Research and Management, 2, pp. 193–208. 2001

C. Kemper et al., Southern right whale (Eubalaena australis) mortalities and human interactions in Australia, 1950-2006, Journal of Cetacean Research and Management, 10 (1), pp. 1–8, 2008

J. J. Meager, Marine wildlife stranding and mortality database annual report 2012. II. Cetacean and Pinniped. Conservation Technical and Data Report, 2, pp. 1–38, 2013

The NOAA tracks unusual mortality events for whales. Looking at data from 2012 to 2024, they said:

"Partial or full necropsy examinations were conducted on approximately half of the whales. Of the whales examined (approximately 90), about 40 percent had evidence of human interaction, either ship strike or entanglement."

NOAA 2016–2024 Humpback Whale Unusual Mortality Event Along the Atlantic Coast

Now that we know what really harms whales, the solution becomes obvious. Scientists have tracked where the whales are and compared their locations to shipping routes in order to redirect ships around high danger zones. Combined with speed restrictions, this should be an effective way to prevent further harm to whale — nothing to do with plastics.

A. C. Nisi et al., Ship collision risk threatens whales across the world's oceans, Science, 386, pp. 870–875, 2024

G. K. Silber et al., The role of the International Maritime Organization in reducing vessel threat to whales: Process, options, action and effectiveness, Marine Policy, 36 (6), pp. 1221–1233, 2012

There is a funny story about the harm to whales. When confronted with the science on what really harms whales, one lady was so desperate to cling to her belief that it must be plastic that she pushed back on the evidence saying, "But I live in Australia and your studies are on whales from other places." This highlights how hard it is to reach people who are already brainwashed. I had to go find studies on whales in her region, which is, of course, nonsensical because whales do not live in one spot. These are the same whales that swim all over the world.

SHARKS AND RAYS

The study used the unusual method of analysing mentions on social media to estimate the harm to sharks and rays.

They found that discarded fishing gear was by far the major problem and also that the effects of marine pollution are dwarfed by the amount of harm from fishing. This is a vital point because accurately evaluating any issue and finding appropriate actions requires putting its impact into perspective.

"The numbers of entangled elasmobranchs reported here are minimal in comparison to the numbers of elasmobranchs caught directly in targeted fisheries or indirectly as bycatch."

K. J. Patton et al., Global review of shark and ray entanglement in anthropogenic marine debris, Endangered Species Research, 39, pp. 173–190, 2019

The science shows that if we want to protect sharks and rays, then the best action is to better regulate the fishing industry.

BIRDS

If online gossip is to be believed, ingestion of plastic is a major threat to birds. For people not gullible enough to believe every headline they see, scientists can offer insights here as well.

"Often, it is difficult to produce evidence for causal links between ingested debris and mortality, and as a consequence, documented cases of death through plastic ingestion are rare. A direct lethal result from ingestion probably does not occur at a frequency relevant at the population level."

M. Bergman, L. Gutow, M. Klages (Eds.), Marine Anthropogenic Litter, Chapter 4, p. 93, Springer 2015

The book says that if there is any effect, it is too small to be relevant. What do other studies say?

"Long-term studies on seabirds have shown that measures to reduce loss of plastics to the environment do have relatively rapid effects. After considerable attention to the massive loss of industrial pellets to the marine environment in the early 1980s, improvements in production and transport methods were reflected in a visible result in the marine environment within one to two decades: several studies from around the globe showed that by the early 2000s the number of industrial granules in seabird stomachs had approximately halved from levels observed in the 1980s."

M. Bergman, L. Gutow, M. Klages (Eds.), Marine Anthropogenic Litter, Chapter 4, p. 105, Springer 2015 Citing Van Franeker & Meijboom, 2002

While certain groups have brought up the topic of plastic pellets (sometimes called "nurdles"), it turns out that this was identified by the US EPA in 1993. Then, action was taken such that the amounts ingested are now far lower and are not increasing, unlike what we have been led to believe.

"Between 1958 and 1959 they found no plastic in prions but from then on there was an upward trend in plastic consumption until 1977. A peak of plastic ingestion was detected in 1985 and 1995 in a number of long-term studies…"

"In contrast to the continuing growth of global plastic use and increase in marine activities, the trend of plastic consumption decreased and stabilized from 2000 onwards approaching the 1980s level."

M. Bergman, L. Gutow, M. Klages (Eds.), Marine Anthropogenic Litter, Chapter 4, p. 85, Springer 2015 Citing Moser & Lee 1992, Robards et al. 1995, Spear et al. 1995, Mrosovsky et al. 2009, Van Franeker et al. 2011, Bond et al. 2013

Is plastic truly the prime culprit when it comes to harm to seabirds? Scientists have investigated that too.

"Obstruction of the gastro-intestinal tract is the leading cause of death. Overall, balloons are the highest-risk debris item; 32 times more likely to result in death than ingesting hard plastic. These findings have significant implications for quantifying seabird mortality due to debris ingestion, and provide identifiable policy targets aimed to reduce mortality for threatened species worldwide."

L. Roman et al., A quantitative analysis linking seabird mortality and marine debris ingestion, Nature - Scientific Reports, 9, 3202, 2019

Again, we find that a focus on plastic is misplaced and that if we want to protect birds, then we should concentrate on rubber balloons.

What is a real threat to birds if it isn't plastic? The top threat is cats. It has been estimated that up to 2 billion birds are killed each year by cats in the USA alone. So, anyone genuinely interested in bird well-being would be better off putting a bell on their cat than fretting about plastic. By the way, bird mortality due to wind turbines is real, but the number of cases is negligible compared to other causes.

S. R. Loss, T. Will & P. P. Marra, Direct Mortality of Birds from Anthropogenic Causes, Annual Review of Ecology, Evolution, and Systematics, 46, pp. 99–120, 2015

W. P. Erikson et al., A Summary and Comparison of Bird Mortality from Anthropogenic Causes with an Emphasis on Collisions, USDA Forest Service Gen. Tech. Rep. PSW-GTR-191 pp. 1029–1024, 2005

OCEAN-BOUND PLASTIC

This is a scheme that claims to prevent plastic from entering the ocean. There is even an Ocean Bound Plastic certification. The idea is that they intercept plastic that would have been washed into the ocean. That sounds like an admirable enough goal, but did you notice how they define "ocean-bound"?

"OBP is an 'Abandoned Plastic Waste' (microplastics, mezzo-plastics and macro-plastics), located within 50km from shores where waste management is inexistent or inefficient. When already located in a landfill or managed dump site, the plastic waste is not considered as OBP. However, when abandoned in an uncontrolled or informal dump site, this waste is considered as OBP."

https://www.obpcert.org/what-is-ocean-bound-plastic-obp/

They used the Jambeck definition, i.e. the definition from a study that has been proven to be invalid because it grossly overestimates plastic getting into the ocean. The reality is that plastic within 50 km (around 35 miles) of the ocean has a much less than 1% chance of ever reaching the ocean. This means that "ocean-bound plastic" was never actually ocean-bound. In short, this is another example of how actions initiated without proper due diligence end up backfiring.

SUMMARY

We have been told that oceans are choking in plastic, that these amounts are huge and increasing exponentially. Harm to turtles, whales, and other marine life is said to be extreme, and we must empty our pockets now to address this urgent emergency.

In stark contrast, comprehensive scientific studies spanning decades and millions of miles of measurements show low amounts of plastic that are not increasing. Consumer items like bags, straws, and bottles are 0.03% of ocean gyre plastic, with no evidence that they are a significant threat.

Studies find that the vast majority of ocean junk is actually discarded fishing nets and other gear.

What is the solution? NGOs are having multiple meetings at the United Nations Environmental Program with the clearly stated goal of reducing plastic consumption and consumer plastics that do not cause harm. They also stated their intention to ignore completely the real, proven danger of discarded nets. This is another example of what happens when certain NGOs are allowed to mislead the public and policymakers. Not only that, but the NGOs work hard to make sure that no independent scientist shows the real data because then their game would be up, their power gone, and the vast income they receive from fiction-mongering would disappear.

The tens of thousands of flights to attend the UNEP INC meetings generate vast impact while the meetings achieve nothing of value. Their net effect is negative (pun intended).

The solution to help the oceans is clear — education, deposits, and fines for the fishing industry.

Ocean clean-up has been proposed and funded, but it is a futile exercise that increases harm because the fossil fuel used and GHG emissions from operating the ships far exceed any potential benefit from cleaning up the tiny amounts of plastic they collect.

DEGRADATION

DEGRADATION

The public's perception is that plastics are bad because they don't degrade. Countless websites, including the WWF, state that it takes 450, 500, or even 1000 years for plastics to degrade. Some even claim that they never degrade; they merely crumble into smaller and smaller pieces. This is from the book that may have started the popular perception of plastics degradation.

"Many plastics take as long as 500 years to decompose. Their very strength and durability make them a persistent pollution problem."

M. Gorman in Environmental Hazards: Marine Pollution, ABC-Clio Inc. USA, 1993

That statement was simply made up, without proof of any kind. Nevertheless, it has been repeated over and over again by groups seeking to demonise plastics. In this chapter, we will look at the current perception and compare it to the scientific evidence. There are thousands of peer-reviewed articles on the topic of plastics degradation. What do they tell us? Is the popular narrative true?

There have been millions of experiments on the degradation of plastics. The reason for that is simple — when a plastic car part, piece of garden furniture, or medical device is made, the manufacturer must be certain that it will last the intended amount of time. What use is a bulletproof Kevlar vest if it crumbles to dust after a week? Plastic pipes bring us clean water. Can you imagine the cost of digging up and replacing those water pipes if they failed after a year or two? Brands want to make high-quality products, and because the cost of premature failure is so high, huge amounts of time and money have been spent researching the degradation of plastics.

Every day we see plastics degrading with our own eyes. Think of the polypropylene garden chairs that become white and brittle until the legs snap off when you sit on them. Think of the polycarbonate car headlamp covers that become yellow and foggy over time.

We are told by the WWF and others that plastic shopping bags take hundreds of years to degrade, but scientists have studied the degradation rate of polyethylene shopping bags, and all peer-reviewed studies found they disintegrate very rapidly, meaning less than one year left outdoors in the open.

"After 9 months exposure in the open-air, all bag materials had disintegrated into fragments."

I. E. Napper, R. C. Thompson, Environmental Deterioration of Biodegradable, Oxo-biodegradable, Compostable, and Conventional Plastic Carrier Bags in the Sea, Soil, and Open-Air Over a 3-Year Period, Environmental Science & Technology, 53 (9), pp 4775–4783, 2019

"This study shows that the real durability of olefin polymers may be much shorter than centuries, as in less than one year the mechanical properties of all samples decreased virtually to zero, as a consequence of severe oxidative degradation..."

T. Ojeda et al., Degradability of linear polyolefins under natural weathering, Polymer Degradation and Stability, 96, pp. 703–707, 2011

Exposure Condition	Degradation Paper Bag	Degradation Plastic Bag
Sunlight	Soft & tearing in 8 - 9 weeks	Transparent & tearing in 10 - 11 weeks
Leaf Pile	Became dry in 10 weeks	Thin, wrinkled with holes in 10 weeks
Soil	Tearing in 7 weeks, pieces after 10 - 12 weeks	Soft & thin after 10 - 12 weeks
Fresh Water	Soft & tearing in 11 - 12 weeks	Thinning after 8 weeks
Salt Water	Soft & tearing in 8 - 9 weeks	Transparent & tearing in 10 - 12 weeks

O. Olaosebikan et al., Environmental Effect on Biodegradability of Plastic and Paper Bags, IOSR Journal of Environmental Science, Toxicology and Food Technology, 8 (1), pp. 22–29, 2014

Unstabilised low density polyethylene (LDPE) lost more than half of its strength in just 30 days when left exposed outdoors and lost over 70% strength in 90 days. The film was seen to crack and tear. Even with stabiliser added, the bags degraded rather rapidly because such items contain low amounts of stabiliser that are rapidly used up. Again, shopping bags are made from LDPE, and NGOs tell us, without evidence, that they take hundreds of years to degrade when science says just the opposite.

M. A. Tuasikal, Influence of Natural and Accelerated Weathering on the Mechanical Properties of Low-Density Polyethylene Films, International Journal of Polymer Analysis & Characterization, 19, pp. 189–203, 2014

Once again, we have been lied to by NGOs who make a living from demonising plastics.

Why do they degrade? Plastics are held together by the same chemical bonds as natural polymers like cellulose, silk, collagen, enzymes, and even the DNA that holds the program responsible for life. Since the chemistry is similar, the degradation rate and final degradation products are similar. All the materials just mentioned degrade to smaller and smaller particles, then to molecules until, eventually, they form carbon dioxide and water. They are attacked by oxygen, heat, and light, and despite what you may have been told, they biodegrade too.

"The ultimate products of degradation are CO_2, H_2O, and biomass under aerobic conditions. Anaerobic microorganisms can also degrade these polymers under anoxic conditions."

J. Arutchelvi et al., Biodegradation of polyethylene and polypropylene, Indian Journal of Biotechnology, 7, pp. 9–22, 2008

Museum curators experience the deterioration of plastic items firsthand. They witness plastic and rubber exhibits becoming brittle and crumbling in real time, and they go to great lengths to preserve the fragile plastic items that reveal our past. I know that because a good friend of mine, Dr. Edward Then, was a plastics conservator at the Victoria & Albert Museum in London, England. As early as 1992, he was charged with working out what plastic each item was made of and how best to preserve it. That is not a simple task because conservators must analyse the exhibits without altering or destroying them, so the techniques they can use are limited to non-invasive types like infrared spectroscopy.

ttp://www.vam.ac.uk/content/journals/conservation-journal/issue-21/plastics-not-in-my-collection/

Books and thousands of peer-reviewed journal articles find that plastics do degrade. That is a scientific certainty — a fact. There is zero doubt. Here are the different ways that plastics are degraded by natural forces.

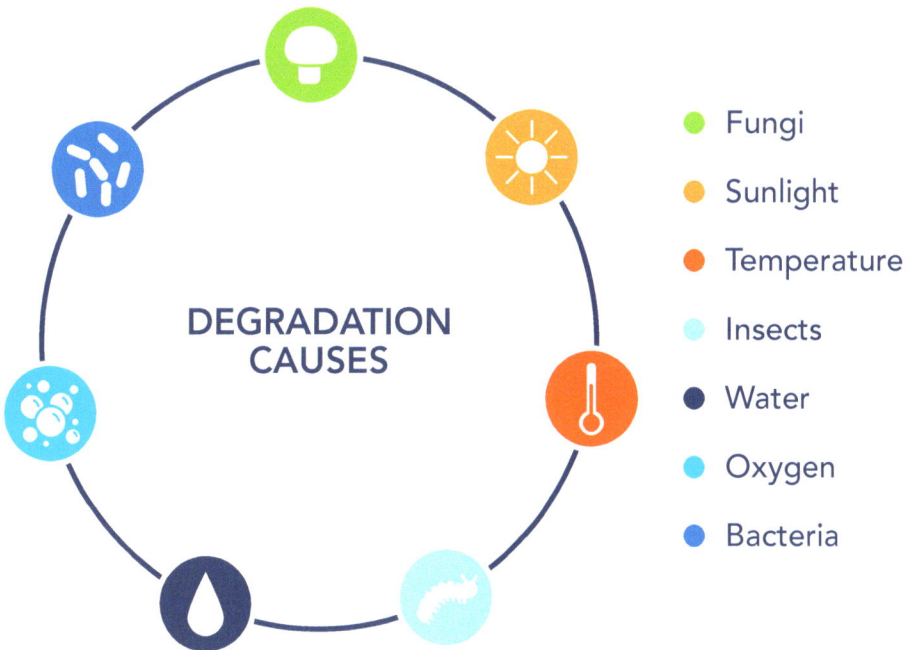

DEGRADATION CAUSES

- Fungi
- Sunlight
- Temperature
- Insects
- Water
- Oxygen
- Bacteria

W. L. Hawkins, Polymer Degradation & Stabilization, Springer Berlin / Heidelberg, 1984

Inamuddin et al. (Eds.), Degradation of Plastic Materials, Materials Research Forum, 2021

Y. Shashoua, Conservation of Plastics: Materials science, degradation and preservation, Routledge, 2008

S. Balasubramanian, Degradation of plastics by Microbes, Lambert Academic Publishing, 2018

M. Srikanth et al., Biodegradation of plastic polymers by fungi: a brief review, Bioresources & Bioprocessing, 9 (42), 2022

G. Weber, U. T. Bornscheuer, R. Wei (Eds.), Enzymatic Plastic Degradation (Methods in Enzymology, Volume 648), AP, 2021

We have firmly established that plastics degrade, rather rapidly in many cases, but do we want them to? Looking at life cycle analyses, the answer is clear — products that are more durable tend to be greener. That being the case, what can we do to make plastics last longer? The answer is to copy Mother Nature. Just like natural nuts and oils contain vitamin E as an antioxidant, the plastics we use contain similar antioxidants and stabilisers. These are added in tiny amounts, usually in the 0–1000 parts per million concentration range, and yet they can greatly extend the useful life of the plastic materials we use. The useful life might be extended from years to decades. You may not realise it, but billions of dollars are spent each year on stabilisers to make plastics last longer and thereby make them greener. Companies would not spend billions on stabilisers for plastics if they really were stable like the NGOs claim.

Polymer Stabilizer Market by Type (Antioxidant, Light Stabilizer, Heat Stabilizer), End-use Industry (Packaging, Automotive, Building & Construction, Consumer Goods), and Region - Global Forecast to 2022 — Markets and Markets Report CH 5459, July 2017

Adding the right stabilisers also helps with recycling. Without any stabiliser, the plastic degrades rapidly and cannot be reused or recycled. Experiments show that an unstabilised polypropylene film degrades and becomes useless in less than a year at room temperature indoors. In fact, PP, one of our greenest and most widely used plastics, would not be of any use at all without a dash of stabiliser.

"Without stabilizers, the degradation of PP is so fast as to make this polymer unsuitable for most purposes. Even at room temperature unstabilized PP loses its mechanical strength within a year."

P. Gjisman, J. Hennekens, J. Vincent, Polymer Degradation and Stability, 39, pp. 271–277, 1993

PVC is another common, versatile, inexpensive, and low environmental impact plastic that requires stabilisers to protect it from degradation when it is melted and processed. However, once properly stabilised, it can remain stable in service for decades.

There are many books filled with studies on the degradation of plastics under all kinds of conditions. Here is one study on the degradation of polyethylene (PE), polypropylene (PP), and polystyrene (PS) plastics outdoors. Degradation is obvious from left to right, even to the untrained eye, as the surface becomes pitted and rougher.

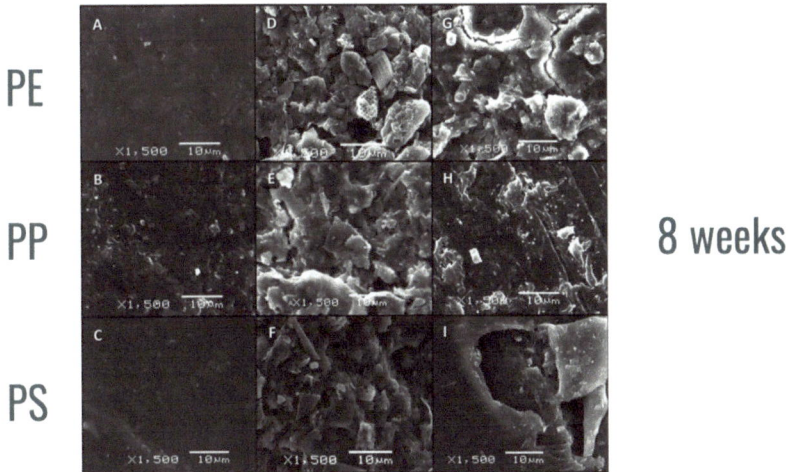

PE
PP   8 weeks
PS

"The results suggest that the degradation of plastic debris proceeds relatively quickly in salt marshes and that surface delamination is the primary mechanism by which microplastic particles are produced in the early stages of degradation."

J. E. Weinstein et al., From Macroplastic to Microplastic: Degradation of High-Density Polyethylene, Polypropylene, and Polystyrene in Salt Marsh Habitat, Environmental Toxicology & Chemistry, 35 (7), pp. 1632–1640, 2016

A REAL-WORLD EXAMPLE

You may have seen me on CBS's *60 Minutes* TV show with Scott Pelley talking about PP medical mesh implanted into people. Polypropylene mesh is used for vaginal repair and for hernias. A class-action lawsuit started when 100,000 women reported problems, and similar lawsuits sprang up about men with hernia mesh. A key topic was the stability of the polypropylene plastic. Such mesh needs around 60 years of stability, but calculations showed it would only last 2–4 years before degrading. The plaintiffs presented evidence that there was not enough stabiliser added, and the wrong kinds of stabiliser were used.

The defence claimed that polypropylene is inert and does not degrade even though massive amounts of peer-reviewed science show the opposite. For example, here is just one study showing that polypropylene degrades through oxidation even at near room temperature.

L. Achimsky et al., On a transition at 80 °C in polypropylene oxidation kinetics, Polymer Degradation and Stability, 58, pp. 283–289, 1997

That was a real-world example of how plastics degrade rapidly and the consequences. We were able to get financial settlements for thousands of women. Note that my role was to show the truth about plastics because, as a professional, independent scientist, my goal is not to promote plastics but rather to expose the facts. My appearance on *60 Minutes* was unpaid, whereas others accepted payment for their work on the show. I worked for free, as I believed it was important for those women to understand the truth about what had been done to them. I later appeared on the BBC and Sky News then assisted in a UK government inquiry, all for free and in the name of justice.

Why are common plastics so sensitive to attack by oxygen, heat, and light? The long molecules that make up plastic materials give strength to the material by tangling together. Only long chains can tangle well, in the same way that only long hair gets tangled. When the polymer molecules are attacked, it only requires the cutting of a few chains for the structure to unravel, leaving the material weak and crumbling. Think of a knitted sweater made of one long piece of yarn. As soon as the yarn is cut, the whole garment can unravel. The same concept applies to the polymer chains that form plastic materials.

DEGRADATION OF OTHER PLASTICS

Polyethylene and polypropylene are chemically similar, and both degrade rapidly. It is only the addition of stabilisers that produces the illusion of stability so that, to the layperson, they appear to be immune to degradation. Together, those two types of thermoplastics make up over 50% of the market, but what about other common plastics? Do they degrade as well?

Another common plastic is PET. Ioakeimidis et al. found that PET bottles degraded, with clear changes in the chemistry found by infrared spectroscopy. After 15 years in the sea, the characteristic chemical bonds were almost gone, indicating severe degradation.

C. Ioakeimidis et al., The degradation potential of PET bottles in the marine environment: An ATR-FTIR based approach, Scientific Reports, 6 (3501), 2016

A more recent study revealed that PET degrades more rapidly than previously thought in ocean water due to the presence of metal ions in the water. 50% degradation (depolymerisation back to the starting materials) was said to occur in 4.5 years and 100% degradation in 72 years.

"According to our research, the time of reaction for a PET conversion of 50% at 35 °C is only 4.5 years in any tropical zone of the Atlantic, Pacific and Indian Oceans or the Caribbean Sea. Also, total PET depolymerization, at a temperature of 30 °C needs only 162 years in any marine water on the globe. All these calculated data provide precise information about the period of depolymerization of waste PET floating in marine waters and correct old estimations of more than 400 years for the total degradation of waste PET."

D. Stanica-Ezeanu & D. Matei, Natural depolymerization of waste poly(ethylene terephthalate) by neutral hydrolysis in marine water, Nature Scientific Reports, 11, 4431, 2021

These numbers do not include the added degradation from ultraviolet light and marine organisms, so actual degradation is likely much faster still.

Although the chemistry of PET degradation is completely different compared to PE and PP, we still see that the plastic degrades over a period of years or decades, not centuries or millennia.

Even polystyrene, usually thought of as very resistant, was found to degrade much more rapidly than previously thought when exposed to sunlight.

"In the current study, we report the first direct evidence of complete oxidation of PS to CO_2 by solar wavebands. All five PS samples were converted to CO_2 by sunlight. For example, when exposing PS to increasing durations of simulated sunlight (up to 72 h), DIC increased, indicating that PS was completely photo-oxidized to CO_2."

C. P. Ward et al., Sunlight Converts Polystyrene to Carbon Dioxide and Dissolved Organic Carbon, Environmental Science & Technology Letters, 6, 11, pp. 669–674, 2019

PVC was also found to be attacked and biodegraded by larvae, thus dispelling the myth that it is impervious.

"The discovery in this study demonstrates that PVC can be depolymerized and biodegraded in Tenebrio Molitor Larvae, which extends observations of PS and PE biodegradation to another major polymer PVC."

B.-Y. Peng et al., Biodegradation of Polyvinyl Chloride (PVC) in Tenebrio molitor (Coleoptera: Tenebrionidae) larvae, Environmental International, 145, 106106, 2020

New York proposed a ban on laundry and detergent pods because they claim such pods do not dissolve or degrade and instead form microplastics. However, the peer-reviewed science shows the opposite.

"In conclusion, PVOH used in liquid detergent capsule films does not meet any of the definitions of

microplastic:(1) it is not micro- or nano-sized; (2) it is highly water-soluble; and (3) it is biodegradable in the environmental conditions where it is discharged."

D. Byrne et al., Biodegradability of Polyvinyl Alcohol Based Film Used for Liquid Detergent Capsules, Tenside Surfactants Detergents, 58 (2), pp. 88–96, 2021

Why are people so keen to propose action before checking the facts first? It is unprofessional and counterproductive.

BIODEGRADATION OF COMMON PLASTICS

When people first realise that common plastics like PE, PP, PVC, and PET degrade, instead of being satisfied and relieved, they instead look for some other reason to cling to their negative attitude. They will say, "Well, perhaps they degrade, but they don't biodegrade." — Or words to that effect. However, they are wrong there too. Conventional plastics do biodegrade. There are many studies from research groups all over the world reporting and measuring the biodegradation of the plastics we use. As this idea is so contrary to the public perception, I will provide plenty of evidence below.

"This review discusses the literature on biodegradation of PE and PP. Most of the examples deal with fungi and bacterial degradation. Pre-treated polymers degrade more easily than the untreated polymers."

J. Arutchelvi et al., Biodegradation of polyethylene and polypropylene, Indian Journal of Biotechnology, 7, pp. 9–22, 2008

"In this study, Lysinibacillus sp., isolated and identified as a novel strain, was investigated to decompose polyethylene and polypropylene. In the microbial cultivation medium without any physicochemical pretreatment, the Lysinibacillus sp. reduced the weight of polypropylene and polyethylene by approximately 4 and 9%, respectively, over 26 days."

J.-M. Jeon et al., Biodegradation of polyethylene and polypropylene by Lysinibacillus species JJY0216 isolated from soil grove, Polymer Degradation and Stability, 191, 109662, 2021

"For LDPE, however, remarkable whitening of the film which was directly in contact with soil was observed. A lot of small holes which are passing through the film was observed around the whitened part. The degradation was more remarkable for samples which were buried in shallow places where the activity of aerobes is high."

The rate of degradation is slower if the plastic is buried but faster if it is first exposed to sunlight to start the degradation process.

"The results show that high-molecular-weight polyethylene can really biodegrade under bioactive circumstances if the test period is long enough."

J.-M. Jeon et al., Biodegradation of low-density polyethylene, polystyrene, polyvinyl chloride, and urea formaldehyde resin buried under soil for over 32 years, Journal of Applied Polymer Science, 56, pp. 1789–1796, 1995

"The Pseudomonas alcaligenes was found to be more effective than Desulfotomaculum nigrificans in degradation of polythene bag at 30 days."

M. Ariba Begum et al., Biodegradation of Polythene Bag using Bacteria Isolated from Soil, International Journal of Current Microbiology and Applied Sciences, 4 (11), pp. 674–680, 2015

Polyethylenes and PVC were also found to biodegrade under marine conditions.

"The mineralization of plastic film was found to be maximum in LDPE followed by HDPE and PVC. Bacterial interaction had increased roughness and deteriorated the surface of plastics which is revealed by the scanning electron microscope and atomic force microscope."

"The results of the present study revealed the ability of marine bacterial strain for instigating their colonization over plastic films and deteriorating the polymeric structure."

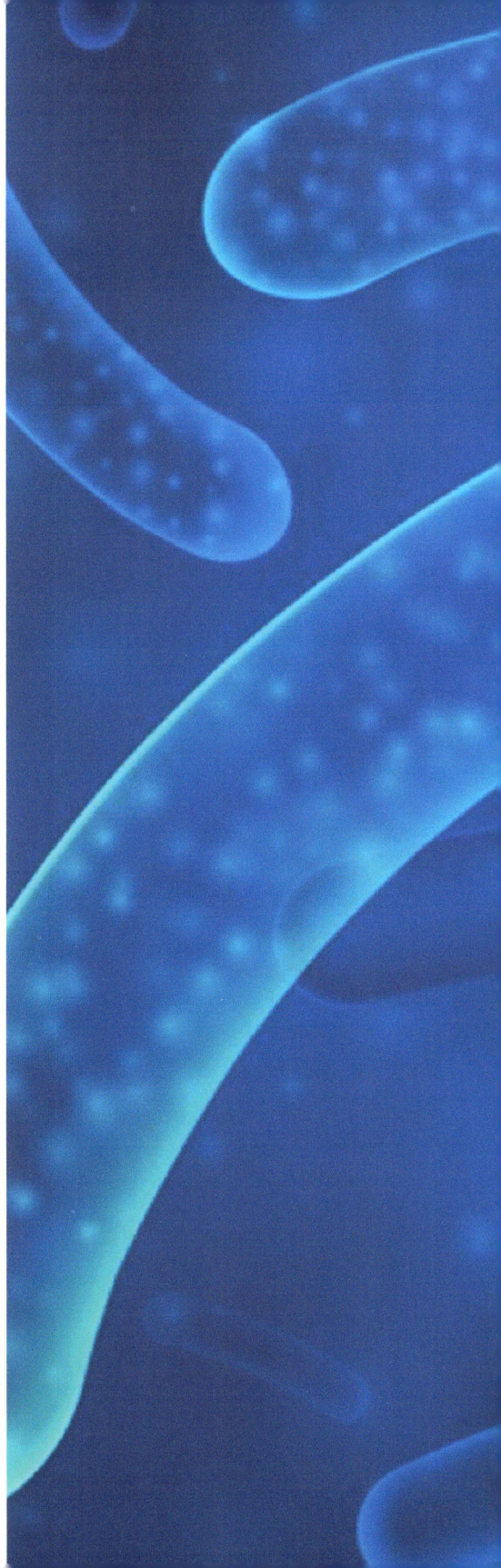

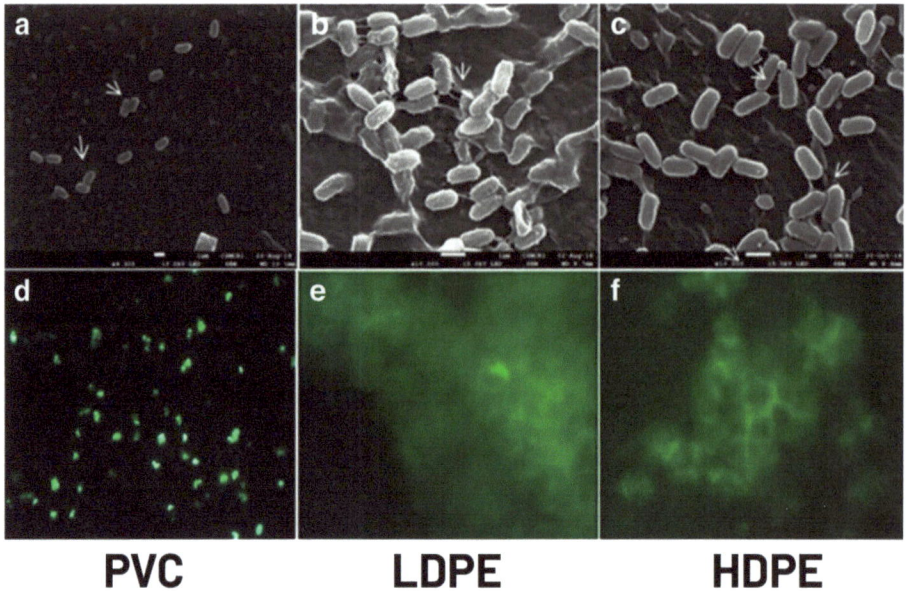

PVC LDPE HDPE

A. Kumari et al., Destabilization of polyethylene and polyvinylchloride structure by marine bacterial strain, Environmental Science and Pollution Research, 26, pp. 1507–1516, 2018

"At least parts of the vast amounts of plastic litter in the ocean may thus serve as a carbon source for fungi and possibly other microbes, too."

A. Vaksmaa et al., Polyethylene degradation & assimilation by the marine yeast Rhodotorula mucilaginosa, ISME Communications, 3 (68), 2023

"This study revealed that the active biodegradation of LDPE film by marine bacteria and these bacteria could reduce plastic pollution in the marine environment."

S. D. Khandare et al., Marine bacterial biodegradation of low-density polyethylene (LDPE) plastic, Biodegradation, 32, pp. 127–143, 2021

People often criticise plastics for not degrading in a landfill, which is unjust because even paper and food degrade slowly in a landfill due to low oxygen levels. Scientists recovered decades-old newspapers that could still be read, which is how they knew how old they were.

W. Rathje & C. Murphy, Rubbish! The Archaeology of Garbage: What Our Garbage Tells Us About Ourselves, Harper Collins, New York, NY, USA 1992

Landfills are designed to slow down degradation because converting solids into carbon dioxide is what most people are trying to avoid. Even so, studies show that PE and PP degrade in a landfill, just like paper and other organic matter do.

"This research analyzed the degradability/biodegradability of polypropylene films (PP) and Bioriented polypropylene (BOPP) polymers after 11 months interred in the São Giácomo landfill in Caxias do Sul."

L. Canopoli et al., Degradation of excavated polyethylene and polypropylene waste from landfill, Science of the Total Environment, 698, 134125, 2020

"SEM and OM revealed the start of degradation/biodegradation processes of the polymeric film in the landfill typified by microorganism colonies on the polymer surface, chromatic alteration and formation of cracks."

C. Longo et al., Degradation Study of Polypropylene (PP) and Bioriented Polypropylene (BOPP) in the Environment, Materials Research, 14(4), pp. 442–448, 2011

"The evidence that biodegradation occurs comes from the increasing concentrations of the methylene chloride extraction products of the incubated polypropylene, together with the contemporary weight loss of the sample. Spectral analysis revealed that the extraction products were mainly hydrocarbons."

"Hence, we suggest that the well-known metabolic flexibility and adaptability of microorganisms and mycelia can result in the biodegradation of isotactic polypropylene and polyethylene, two macromolecules that supposedly are highly recalcitrant to biological metabolism."

I. Cacciari et al., Isotactic polypropylene biodegradation by a microbial community: physicochemical characterization of metabolites produced, Applied and Environmental Microbiology, 59 (11), pp. 3695–3700, 1993

PET was found to degrade in sunlight and even more quickly when moisture and soil were present as well. Polymer chain scissions means breaking the long molecules into shorter ones. Such degradation weakens the plastic material.

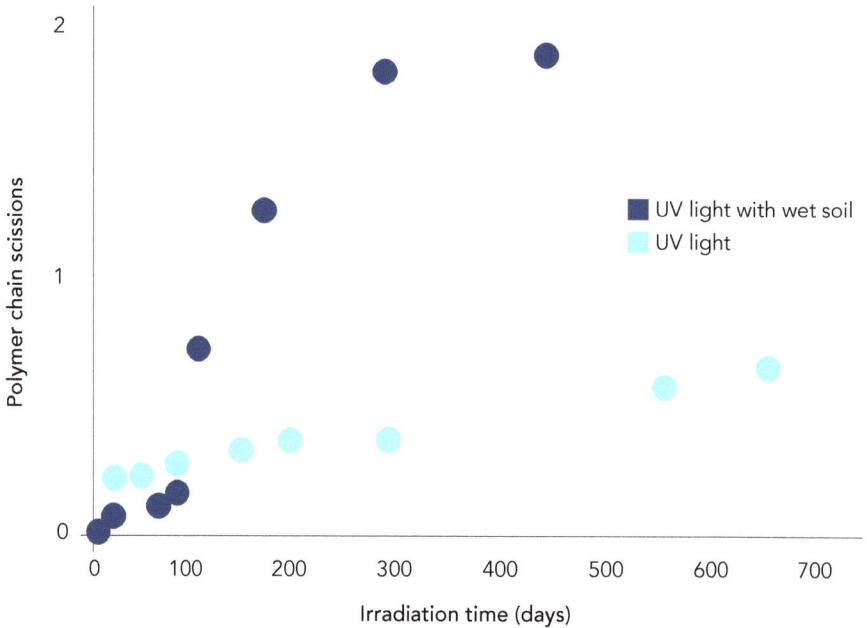

N. Allen et al., Physicochemical aspects of the environmental degradation of poly(ethylene terephthalate), Polymer Degradation and Stability, 43, pp. 229–237, 1994

"FTIR analysis implies structural changes in biodegraded PET samples unlike the control. The biodegradation is further substantiated by SEM which manifested the development of fissures and a sign of significant erosions which were progressive with the incubation time."

M. G. H., Zaidi, Comparative in situ PET biodegradation assay using indigenously developed consortia, International Journal of Environment and Waste Management, 13 (4), pp. 348–361, 2014

"We eventually found a unique microbial consortium, named No. 46, in a landfill. This consortium is able to grow on low-crystallinity PET film; it assembles on the film and utilizes PET as a major carbon and energy source, degrading it into CO_2 and water."

K. Hiraga et al., Biodegradation of waste PET, Science & Society, 20, e49365, 2019

We have now seen a robust array of studies illustrating that PE, PP, and PET biodegrade, but what about polystyrene? Most people believe it to be non-degradable.

"Academics researchers and "citizen scientists" from 22 countries confirmed that yellow mealworms, the larvae of Tenebrio molitor Linnaeus, can survive by eating polystyrene (PS) foam."

"The results indicate that mealworms from diverse locations eat and metabolize PS and support the hypothesis that this capacity is independent of the geographic origin of the mealworms, and is likely ubiquitous to members of this species."

S.-S. Yang et al., Ubiquity of polystyrene digestion and biodegradation within yellow mealworms, larvae of Tenebrio molitor Linnaeus (Coleoptera: Tenebrionidae), Chemosphere, 212, pp. 262–271, 2018

"Fed with Styrofoam as the sole diet, the larvae lived as well as those fed with a normal diet (bran) over a period of 1 month."

"Within a 16 day test period, 47.7 % of the ingested Styrofoam carbon was converted into CO_2."

"The discovery of the rapid biodegradation of PS in the larval gut reveals a new fate for plastic waste in the environment."

Y. Yang et al., Biodegradation and Mineralization of Polystyrene by Plastic-Eating Mealworms: Part 1. Chemical and Physical Characterization and Isotopic Tests, Environmental Science & Technology, 49, 20, pp. 12080–12086, 2015

You read that correctly — mealworms fed only with polystyrene foam survived perfectly for a month and converted the plastic fully into carbon dioxide. I was surprised too. In fact, I was so surprised that I checked to make sure this was real and replicated by other research groups.

The same mealworms could also eat other plastics, including polyethylene and polyurethane.

"Microbial degradation in environmental conditions in vitro is extremely slow for major plastics at degradation rates on the basis of a month or even a year time, but recent discoveries show that the fast biodegradation of specific plastics, such as PS, PE, and PUR, in some invertebrates, especially insects, could be enhanced at rates on basis of hours."

X.-G. Yang et al., Plastic biodegradation by in vitro environmental microorganisms and in vivo gut microorganisms of insects, Frontiers in Microbiology, 13, 1001750, 2023

It is not only one type of mealworm that can perform this amazing feat; other larvae and also snails can do the same.

"For the first time, this study reveals that land snails Achatina fulica has the capacity to depolymerize and biodegrade polystyrene. Mass balance, GPC, FTIR and 1H NMR analyses confirmed the limited extent de-polymerization and oxidation of PS polymers, which supported the occurrence of biodegradation."

"Concerning land snail was one of the mostly popular and rapidly proliferated terrestrial animals, these findings are significant in regards to the fate of plastic litter and its biodegradation in soil environments."

Y. Song et al., Biodegradation and disintegration of expanded polystyrene by land snails, Science of the Total Environment 746, 141289, 2020

So, insects and snails can biodegrade plastic, and it turns out that bacteria can degrade a wide range of plastics as well.

"This review has discussed the microorganisms and enzymes reported to biodegrade these synthetic polymers. Many strains of Pseudomonas and Bacillus have been observed to degrade complex, recalcitrant compounds such as polyaromatic hydrocarbons, and have been associated with the

partial degradation of a wide-range of petro-plastics, including PE, PS, PP, PVC, PET and ester-based PU. The gut microbes in insects have also been found to depolymerize PE, PS and PVC polymers. Enzymes specifically associated with depolymerization of PET and ester-based PU have been identified and intensively studied, while enzymes that effectively depolymerize PE, PP, PS, and PVC have not yet been identified and characterized."

N. Mohanan et al., Microbial and Enzymatic Degradation of Synthetic Plastics, Frontiers in Microbiology, 11, 580709, 2020

"After considering the above results of the present study, it is to be concluded that PET and PS can be degraded by micro-organisms (biodegradation) like Pseudomonas aeruginosa, Bacillus subtilis, Staphylococcus aureus, Streptococcus pyogenes, and Aspergillus niger, present in different types of soils."

K. Asmita et al., Isolation of Plastic Degrading Micro-organisms from Soil Samples Collected at Various Locations in Mumbai, India, International Research Journal of Environment Sciences, 4 (3), pp. 77–85, 2015

Not only do plastics degrade by heat, light, and oxygen and biodegrade via bacteria and insects, but fungi are also proven to contribute to plastics biodegradation.

"The oxidation or hydrolysis by the enzyme creates functional groups that improve the hydrophilicity

of polymers, and consequently degrade the high molecular weight polymer into low molecular weight. This leads to the degradation of plastics within a few days. Some well-known species which show effective degradation on plastics are *Aspergillus nidulans, Aspergillus flavus, Aspergillus glaucus, Aspergillus oryzae, Aspergillus nomius, Penicillium griseofulvum, Bjerkandera adusta, Phanerochaete chrysosporium, Cladosporium cladosporioides, etc.*, and some other saprotrophic fungi, such as *Pleurotus abalones, Pleurotus ostreatus, Agaricus bisporus* and *Pleurotus eryngii* which also helps in degradation of plastics by growing on them."

PE: *Phanerochaete chrysosporium, Aspergillus, Cladosporium, Fusarium, Penicillium, Phanerochaete, Pencilli-um. Simplicissimum, Aspergillus niger, Aspergillus japonicas and Fusarium. sp., Penicillium chrysogenum NS10*

PP: *Bjerkandera adusta, Lasiodiplodia theobromae, Coriolus versicolor*

PS: *Cephalosporium spp., Mucor spp. Gloeophyllum striatum, Gloeophyllum trabeum DSM 1398, Pleurotus ostreatus, Phanerochaete chrysosporium*

PUR: *Gliocladium roseum, Aspergillus spp., Emericella spp., Fusarium spp., Penicillium spp., Trichoderma spp., Gliocladium pannorum, Nectria gliocladiodes, Penicillium ochrochloron, Aureobasidium pullulans, Rhodotorula aurantiaca, Kluyvermyces spp.*

PC: *Phanerochaete chrysosporium NCIM 1170, Geotrichum spp., Fusarium, Ulocladium, Chrysosporium, Penicillium*

PET: *Fusarium, Humicola, Candida antarctica, Aspergillus sp., Penicillium sp., Fusarium sp.*

PVC: *Cochliobolus sp., Phanerochaete chrysosporium, Aspergillus niger, Penicillium funiculosum ATCC 9644, Trichoderma viride ATCC 13631, Paecilomyces variotii CBS 62866, Aureobasidium pullulans, Chaetomium globosum, Rhodotorula aurantiaca, Kluyveromyces spp.*

M. Srikanth et al., Biodegradation of plastic polymers by fungi: a brief review, Bioresources & Bioprocessing, 9 (42), 2022

The scientific evidence is clear — plastics degrade and biodegrade.

BIODEGRADABLE PLASTICS

Biodegradable plastics are designed to degrade, but do they make sense? They seem superficially attractive because we could throw our litter on the ground or in the ocean and then "abracadabra," it would vanish all by itself. That sounds marvellous, doesn't it? However, life cycle analysis (LCA) studies show that biodegradable plastics have a greater impact than normal plastics like PE and PP. Of course, they are more expensive too and they have worse properties. Plus, when they degrade, they rapidly release carbon dioxide, which is just what people are campaigning against because it is a greenhouse gas.

What about the particles formed when conventional plastics degrade? One common argument proposed for biodegradable plastics is to prevent microplastics. Some people believe plastics degrade to small particles and then degradation stops, but that is not the case. In fact, the smaller the plastic pieces are, the faster they degrade because oxygen and bacteria can attack them more readily. The reason is simple — degradation occurs mainly at the surface and the smaller the particles become, the greater the surface area exposed. More about that later.

Oxo-degradable plastics are where a catalyst (usually iron, nickel, manganese, or cobalt stearate) is added to a plastic like PE or PP to make it break down more rapidly. They are sold as green products, but the green claims do not stand up to scrutiny. Firstly, we know that durable products create less impact, so speeding up failure is unwise. Degradation means converting solids into greenhouse gas, which is the opposite of what most people consider desirable. In addition, the catalysts can contaminate the recycling stream, destabilising the rest of the PE and PP plastic and ruining their recyclability. We know PE and PP degrade rather rapidly in the environment anyway, so if we wanted those plastics to degrade faster, there is no need to add a catalyst. Instead, it would be cheaper to just remove the stabiliser. So, for good reason, oxo-degradables have been banned in the EU, and other regions are likely to follow.

We are led to believe that plastics are intrinsically evil because they last forever, whereas other materials do not. Is that really the case? No, it is not, because other common materials like ceramics, metals, stone, and glass all take longer to degrade than plastics do. Even paper can take longer to degrade than common plastics like PE, PP, and PET, depending on the conditions. The oldest paper documents known are over 1000 years old and still readable. In fact, it has been estimated that paper takes 2700 years to degrade at room temperature when dry. Compare that to polypropylene film, which has been shown to disintegrate in under one year. Why does paper take such a long time to fragment and decay? The answer is that paper contains a large amount of natural stabiliser called "lignin," which is very effective at protecting against oxidation.

As a rule of thumb, a piece of common plastic like PE or PP will degrade at about the same rate as another piece of organic matter of the same size and shape. So, a PE or PP film will degrade similarly to a leaf or a piece of paper. Why is that? It is because PE and PP are organic materials made of carbon-carbon bonds, just like other substances such as cellulose, lignin, cotton, and so on.

When the object is thicker, degradation takes far longer. Fallen sequoia trees have remained intact for at least 500 years with hardly any degradation (Scott, 1999) in the same way that a gigantic piece of plastic, metal, or glass would take much longer to degrade.

Gerald Scott, Polymers and the Environment, RSC Paperbacks, p. 97, 1999

The exact degradation rate depends on temperature, the size of the object, the amount of sunlight, and so on, but the fact remains that common plastics degrade as quickly or even more quickly than the other materials we encounter.

Clearly, claiming plastics are bad because they take longer to degrade than other materials is not a valid argument, as it is not true.

It has been claimed that plastics create a problem because they eventually release CO_2 when they degrade. Would that be a fair criticism of plastics relative to other materials? The answer is no because all organic matter does that too — leaves, wood, cotton, jute, hemp, and paper all degrade in the same way.

PLASTIC PREJUDICE

When we discover a 400-year-old wooden ship in the ocean, we celebrate, build a museum, and sell tickets to look at this "treasure." The same applies when we find 2000-year-old Roman coins made of metal. Stonehenge, a bunch of 5000-year-old rocks, attracts a million visitors per year while 15 million flock to see the pyramids.

Whether it is glass, clay, stone, animal remains, wood, or metal, we are filled with joy to find it, and the older it is, the better. A recent scientific paper even hailed the discovery and analysis of 2700-year-old human excrement.

F. Maixner et al., Hallstatt miners consumed blue cheese and beer during the Iron Age and retained a non-Westernized gut microbiome until the Baroque period, Current Biology 31, pp. 1–14, 2021

There is a clear plastics prejudice at work, whereby it is implied that plastics are evil if they take a long time to degrade when every other material is celebrated when it does not degrade. How unjust.

SUMMARY

Let us summarise what we have discovered and what policies might make sense based on the evidence. We have seen that the notion that plastics don't degrade is false and is, therefore, not a fair or valid criticism.

We know from life cycle analyses that the degradation of plastics is not desirable because it makes products less green. Durable products usually minimise environmental impact. Therefore, in most cases, we want to increase the life of plastics, and we do that by adding the appropriate type and amount of stabiliser.

It follows that biodegradable plastics make little sense. They increase harm to the environment, according to LCA studies, in part because they rapidly release carbon dioxide as they degrade. They also cost more and have worse properties than standard plastics we are familiar with.

Biodegradable plastics are not a solution to litter either. Quite the opposite: degradables exacerbate the problem because they encourage people to litter more.

It turns out that the greenest path is to continue using the plastics that cause the least impact, such as PE, PP, PVC, and PET. Stabilisers should be added to adjust the degradation rate and to ensure that the material is in a fit condition to be recycled into new objects.

Thin PE shopping bags contain minimal amounts of stabiliser and disintegrate in under one year outdoors, which is a similar rate to paper bags. At the other end of the scale, we have durable products like water pipes, which are thicker with more and better stabilisers added to ensure that they last a hundred years or more.

This is the ideal situation in which we can control the plastics degradation rate to be optimal for each different use case.

Knowing all of this, it becomes clear that people are not really against plastics because they do not degrade rapidly enough for them; after all, they do not care that other materials like concrete, metal, glass, and ceramics all degrade slower than plastic. Nor do people criticise paper and wood, even though they degrade at a similar speed to a similarly sized piece of PE or PP plastic. No, the real reason that people want plastic to degrade faster is so that they can drop it on the floor and have it magically vanish. This is the only explanation that makes sense. It is the driving force behind sales of biodegradable and compostable plastics.

This is a rare instance where scientists know the answers, but it might be better not to communicate them too widely to the public because when the customer thinks that the litter will degrade, then they litter more.

TOXICITY, ADDITIVES & MICROPLASTICS

TOXICITY, ADDITIVES & MICROPLASTICS

There is a natural tendency for people to equate natural with safe and synthetic with dangerous, but there is no scientific basis for that sentiment. In fact, 4 out of the 5 most toxic chemicals are natural.

Material	Source	Toxicity LD_{50} (mg/kg = ppm)
Ricin	Castor beans	1 - 20
VX	Synthetic	3
Batrachotoxin	Frogs	2
Maitotoxin	Plankton	0.2
Botulinum	Bacteria	1×10^{-6}

S. Cotton, Handle with care — the world's five deadliest poisons, University of Birmingham Chemistry Department, The Conversation, 2016

This misconception may be one reason people assume synthetic materials like plastic must be toxic, even in the absence of evidence to support the idea.

ARE PLASTICS TOXIC?

No, they are not. Are you convinced yet? Let's take a look at the science.

On 12 May 2021, the *National Observer* reported on the Canadian federal government's announcement about the addition of "plastic manufactured items" as toxic substances under Schedule 1 of the Canadian Environmental Protection Act (CEPA). The minister responsible was Steven Guilbeault. That was quite an outlandish and arrogant decision given that we have decades of data to the contrary and that regulatory bodies the world over have approved plastic utensils, food containers, blood bags, and implantable devices (hips, knees, pacemakers). Yet somehow this former Greenpeacer feels that he knows better than every scientist in the world. It reminded me of the story of King Canute.

"Canute set his throne by the seashore and commanded the incoming tide to halt and not to wet his feet and robes. Yet continuing to rise as usual [the tide] dashed over his feet and legs without respect to his royal person."

On 16 November 2023, a Canadian Federal Court justice overturned the federal government's 2021 declaration that labelled all plastic items as toxic, citing overreach under the Environmental Protection Act. The justice stated that the evidence "has not shown that there is a reasonable apprehension of harm for every plastic manufactured item."

And so it is, that no matter how powerful or arrogant a person may be, their declarations do not stand up in the face of reality. Politicians do not define what is toxic; only scientists can do that, and they have indeed studied the subject in detail. The older method you may have heard of is the LD_{50}, which is the dose fed to a rat or mouse that kills 50% of the test group. It is a measure of acute toxicity. The longer-term, so-called "chronic" toxicity is measured by feeding the test animals for weeks or months to determine the maximum amount that can be ingested repeatedly with no observable effect. This is known as the no-observed-adverse-effect level, or NOAEL for short. How do plastics compare to other substances we routinely come into contact with?

Material/Substance	Acute Toxicity LD_{50} 70 Kg Person	Chronic Toxicity NOAEL 70 Kg Person
Sugar	~ 2000 g	60 g
Alcohol (ethanol)	~ 500 g	~ 12 g
Polyethylene	> 350 g	> 50 g
Polypropylene	> 350 g	> 140 g
Polyethylene Terephthalate	> 350 g	N/A
PVC (unplasticized)	> 350 g	N/A
PTFE – Teflon®	> 140 g	> 140 g
Calcium Carbonate - Filler	> 350 g	> 70 g
Talc Mineral - Filler	> 350 g	> 60 g
Calcium Stearate - Lubricant	> 700 g	> 140 g
Irgaphos® 168 - Stabilizer	> 350 g	140 g
Irgaphos® 1076 - Stabilizer	> 350 g	70 g
Irgaphos® 1010 - Stabilizer	> 350 g	100 - 200 g
Irgaphos® 1330 - Stabilizer	> 350 g	35 g
Iron Powder	> 350 g	14 g
Table Salt	~ 200 g	4 g
Caffein	14 g	0,25 g
Copper Metal	30 g	~ 1 g
Cu Dissolved	3.5 g	0.005 g
Nicotine	0.7 g	0.00006 g

As we can see, plastic materials are "non-toxic" and some of the safest materials we have. You could eat a cup of plastic pellets every day for months, and nothing would happen. The US EPA's Toxic Substances Control Act even created the so-called "polymer exemption" in recognition of the exceptional safety of plastics compared to other classes of substances. Since polymer molecules are so large, they cannot migrate (move around), which makes them intrinsically safer than small molecules.

The NOAEL levels were determined using ingestion of plastic pellets that are defined as microplastics, meaning they are 5 mm or smaller. This means that the toxicity — or should I say, the non–toxicity — of microplastics has been well-established experimentally for many years.

When animals ingest plastic, it passes right through without any effect, according to several studies, such as this one on PVC, nylon, UHMWPE, PS, MDPE, and fish.

"In conclusion, the dietary exposure of S. aurata to 6 common types of virgin microplastics did not induce stress, alter the growth rate, cause pathology, or cause the microplastics to accumulate in the gastrointestinal tract of the fish."

B. Jovanović, Virgin microplastics are not causing imminent harm to fish after dietary exposure, Marine Pollution Bulletin, 130, pp. 123–131, 2018

That study also highlighted other studies that had not been performed properly, thus creating unjust concern. We will see later that this is a recurring theme. Researchers conduct experiments in ways designed to produce scary but invalid results.

"However, in previous experimental setups, fish were usually exposed to unrealistically high concentrations of microplastics, or the microplastics were deliberately contaminated with persistent organic chemicals; also, in many experiments, the fish were exposed only during the larval stages."

Here is another study confirming no effect from microplastics when experiments are done properly under realistic conditions.

"However, after one month of detoxification, no MPs were found in the gastrointestinal tracts of fish, reflecting no long-term retention of MPs in Sparus aurata digestive system. According to results from this study, exposure of fish to MP enriched diets does not affect fish size neither the Fulton's condition index as both parameters increased with time in all treatments (control, virgin and weathered)"

C. Alomar et al., Microplastic ingestion in reared aquaculture fish: Biological responses to low-density polyethylene controlled diets in Sparus aurata, Environmental Pollution, 280 (1), 2021

And another two:

"No mortality occurred during the feeding trial and there were no apparent signs of significant distress or adverse effects on the fish. We found no significant differences in growth performance…"

"No accumulation of HDPE was detected in fish collected 24 h post-feeding…"

X. Lu et al., Chronic exposure to high-density polyethylene microplastic through feeding alters the nutrient metabolism of juvenile yellow perch (*Perca flavescens*), Animal Nutrition, 9, 2022

C. Alomar et al., Microplastic ingestion in reared aquaculture fish: Biological responses to low-density polyethylene controlled diets in Sparus aurata, Environmental Pollution, 280 (1), 2021

We are told that plastic pellets (nurdles) and microplastics poison and accumulate in fish, but science shows the opposite. In reality, they pass right through.

There was a study claiming harm, but it was retracted after those scientists were reported for manipulation of data, which led to an investigation.

O. Lönnstedt & P. Eklöv, Environmentally relevant concentrations of microplastic particles influence larval fish ecology, Science, 352, pp. 1213–1216, 2016

There has been a lot of talk about phthalates, which are used to soften a small portion of plastics. They are not used in polyethylene, polypropylene, PET, polycarbonate, polystyrene, ABS, or most plastics you use. Phthalates are not used in PVC pipes because they are made of rigid, unplasticised PVC. Rather, they are used to soften some PVC products that need to be soft. There have been comprehensive studies spanning decades. We know that there is no cause for alarm because exposure due to plastics and various other sources is extremely low.

This review of the science around phthalates found no reason for concern, agreeing with the FDA's position.

"Analysis of all of the available data leads to the conclusion that the risks are low, even lower than originally thought, and that there is no convincing evidence of adverse effects on humans. Since the scientific evidence strongly suggests that risks to humans are low, phthalate regulations that have been enacted are unlikely to lead to any marked improvement in public health."

M. A. Kamrin, Phthalate risks, phthalate regulation, and public health - a review, Journal of Toxicology and Environmental Health, Part B, 12, pp. 157–174, 2009

Exposure is far higher for workers in PVC plants, as one would expect. However, high exposure is also found for massage therapists, nail and beauty salon employees, perfume sales-

people, and people taking certain medications containing phthalates. No one talks about those other exposure sources, perhaps because their real interest is not in phthalates but in attacking plastics. In the case of perfume, you literally spray phthalate right on the skin, which is far worse than holding a piece of plasticised PVC because, in the latter case, the additive only comes out of the plastic very slowly.

P.-C. Huang et al., Characterization of phthalates exposure and risk for cosmetics and perfume sales clerks, Environmental Pollution, 233, pp. 577–587, 2018

BPA is similar in that there are decades of studies and an agreement that exposure from plastics and all other BPA sources is far below recognised safe limits.

"In general, the total exposure to BPA is several orders of magnitude lower than the current tolerable daily intake of 50 µg/kg bw/ day."

T. Geens et al., A review of dietary and non-dietary exposure to bisphenol-A, Food and Chemical Toxicology, 50, pp. 3725–3740, 2012

BPA can form at extremely low concentrations when polycarbonate plastic is left in contact with water, but the amounts are too low to present a problem.

"BPA was only detected in a sample from a polycarbonate container at levels well below the EFSA total daily intake value."

C. Rowell et al., Is container type the biggest predictor of trace element and BPA leaching from drinking water bottles?, Food Chemistry, 202, pp. 88–93, 2016

Other sources of BPA, like thermal paper, are much more of a problem, but there is little mention of that in the press, presumably because the concern is not really about BPA but more about finding ways to unjustly demonise plastics.

Having established that plastics and additives for plastics are not toxic, the next topic is microplastics. There is a perception that plastic particles are a new, previously unrecognised threat to humanity and animals. Is that really the case?

ARE PARTICLES DANGEROUS?

Probably the first question to address is whether we need to be concerned about the health effects of particles in general. The short answer is that yes, particles can and do cause serious health effects, but as with any topic, there is a little more to it than that. The threat level depends on the type of particle, the size, and the dose.

Fine particles under 10 microns and especially under 2.5 microns in size can cause health problems. A review article stated:

"The World Health Organization (WHO) estimated that in the year 2012, ambient air pollution was responsible for 3.7 million annual deaths (which represents 6.7 % of the total deaths), causing worldwide 16 % of deaths for lung cancer, 11 % for chronic obstructive pulmonary disease, more than 20 % for ischemic heart disease and stroke and 13 % for respiratory infection."

P. M. Mannucci et al., Effects on health of air pollution: a narrative review, Internal & Emergency Medicine, 10 (6), pp. 657–62, September 2015

"Nine out of 10 people breathe air that does not meet World Health Organization pollution limits. Air pollutants include gasses and particulate matter and collectively are responsible for ~8 million annual deaths. Particulate matter is the most dangerous form of air pollution, causing inflammatory and oxidative tissue damage."

J. T. Pryor et al., The Physiological Effects of Air Pollution: Particulate Matter, Physiology and Disease, Frontiers in Public Health, 10, (82569), 2022

In areas with heavy pollution, health problems exist, but when particle concentrations are lower, the body's natural defence system can cope. Think of it like a castle wall. If a few invaders try their luck, then they are easily repulsed. But if an onslaught of millions were to try, then they would overrun the castle walls. This makes it plain why dose is so important in the field of toxicology. What may be benign or even beneficial at low concentrations will almost certainly become a problem at extremely high doses.

Given that particles can indeed pose a threat, is there reason to be especially concerned about plastic particles? What are the concentrations and are they toxic? We have all heard the scare stories, but remember that in every chapter so far, science has completely contradicted the message that the public hears. Since this is such an emotive topic, I have read over 500 studies on this one subject, which may well be the most in-depth, independent, and unfunded review on microplastics. That was a painful experience for me, but the good news is that scientists already have all the answers.

DUST PARTICLES & HEALTH

While we are talking about dust, what is in it and how dangerous is the plastic in it compared to the many other types of particles?

Material	Safety
Polyethylene PE	Non-toxic
Polypropylene PP	Non-toxic
Polyester PET	Non-toxic
Cement dust	Respiratory effects
Coffee dust	Respiratory effects
Metal dust	Respiratory effects
Quartz dust	Carcinogenic
Wood dust	Carcinogenic
Leather dust	Carcinogenic
Soot dust	Carcinogenic

J. A. Styles & J. Wilson, Comparison between in vitro toxicity of polymer and mineral dusts and their fibrogenicity, The Annals of Occupational Hygiene, 16 (3), pp. 241–250, November 1973

IARC Monographs Volume 100 A Review of Human Carcinogens, World Health Organization, 2012

Note that I cited a study from 1973, over 50 years ago, just to highlight the point that this is not some new, previously unrecognised topic. Quite the reverse, in fact — we have decades of testing right up to the present on dust and plastic particles.

While plastics are found to be safe, what may surprise many is just how dangerous some of the other particles are. Quartz is one of the most common rocks. When we go to the beach, we merrily bathe in the sunlight, which can give us cancer while breathing in quartz dust, which can also give us cancer. Workers are exposed to dangerous levels of quartz, including those in factories, sawing quartz countertops, and even farmers ploughing the fields can be exposed to levels above recognised safe limits.

"Twelve of 138 respirable dust measurements (9%) and 18 of 138 respirable quartz measurements (13%) exceeded commonly used occupational exposure limits of 2 mg^{-3} and 100 µg m^{-3}, respectively. The highest time weighted average respirable quartz concentration of 626 µg m^{-3} was during wheat planting activities. Fifty-seven percent of the respirable quartz measurements exceeded the American Conference of Governmental Industrial Hygienists (ACGIH) Threshold Limit Value (TLV) of 25 µg m^{-3}. Quartz percentages of the respirable dust ranged from 0.3 to 94.4% with an overall median of 13.4%."

A. J. Swanepoel et al., Quartz exposure in agriculture: literature review and South African Survey, Annals of Occupational Hygiene, 54 (3), pp. 281–292, 2010

"China appears to have the highest burden of silicosis, with more than 500,000 cases recorded between 1991 and 1995, and 6000 new cases and more than 24,000 deaths reported annually."

K. Steenland & E. Ward, Silica: A Lung Carcinogen, CA: A Cancer Journal for Clinicians, 64, pp. 63–69, 2014

"Wood dust was classified as carcinogenic to humans."

"Strong and consistent associations with cancers of the paranasal sinuses and nasal cavity have been observed both in studies of people whose occupations were associated with wood-dust exposure and in studies that directly estimated wood-dust exposure."

IARC Monographs on the Evaluation of Carcinogenic Risks to Humans Volume 62
https://www.cancer.gov/about-cancer/causes-prevention/risk/substances/wood-dust

S. D. Stellman et al., Cancer Mortality and Wood Dust Exposure Among Participants in the American Cancer Society Cancer Prevention Study-II (CPS-II), Journal of Industrial Medicine, 34, pp. 229–237, 1998

Somehow, there seems to be very little concern over the proven dangerous particles, like quartz and wood dust, that cause health problems and even mortalities.

So, what about the plastic fraction of the dust we breathe? How many deaths have been attributed to that? The answer is zero. The portion of dust particles deemed as respirable is below 10 microns in size, abbreviated PM_{10}, and the plastics contribution to that is negligible.

"Therefore, the [microplastic] MP concentration in the air has a negligible contribution to the PM_{10} levels, even at the 95th percentile."

Nur Hazimah and Mohamed Nor, Lifetime Accumulation of Microplastic in Children and Adults, Environmental Science & Technology Journal, 55 (8), pp. 5084–5096, 2021

A comprehensive breakdown of the troublesome fraction of dust below 10 microns in particle size (PM_{10}) globally highlighted the sources of concern for health. Plastics were not even mentioned as a problem.

A. Mukherjee & M. Agrawal, World air particulate matter: sources, distribution and health effects, Environmental Chemistry Letters, 15, pp. 283–309, 2017

It is very clear where PM_{10} dust comes from, and it is not plastics. Although there are actual sources of toxic and carcinogenic (cancer-causing) dust listed in the table, no one appears interested in discussing those genuine problems.

This information highlights the double standard applied to plastics. We are happy to ignore real, proven dangers completely and instead obsess over imaginary ones. As a scientist, I prefer to worry over what deserves my attention and not spend time and money on matters that don't matter.

What about indoor particles and the contribution from plastic? Once again, the plastic fraction of dust is so low that it is not even mentioned in most studies. Other sources of particles, such as skin particles, quartz, spores, and even cooking, dominate the scene.

"The highest mean number concentrations were due to complex cooking, producing average number concentrations of 35,000–50,000 cm^{-3}, compared to 12,000 cm^{-3} outdoors and less than 3500 cm^{-3} indoors when no sources were observed. A strong contribution of the vented gas-powered clothes dryer was also noted (30,000 cm^{-3})."

L. Wallace, Indoor Sources of Ultrafine and Accumulation Mode Particles: Size Distributions, Size-Resolved Concentrations, and Source Strengths, Aerosol Science & Technology, 40, pp. 348–360, 2006

For comparison, the concentration of plastic in indoor dust was under 0.001 cm^{-3}.

M. A. Bhat, Airborne microplastic contamination across diverse university indoor environments: A comprehensive ambient analysis, Air Quality, Atmosphere & Health, 9, 2024

Yet again, we find that the focus on plastic particles is not justified by the evidence.

MICROPLASTICS EXPOSURE

We have been told that we eat a credit card of plastic per week. The WWF tells us that based on a study they paid for. Other non-profits and the media repeated the claim. When considering evidence, it is always best to check other sources of information, preferably impartial ones.

So, what does the best impartial scientific study have to say about microplastic ingestion by humans? The authors of that study specifically state that the WWF study is wrong; in fact, it is so wrong due to a "mistake" that one can hardly believe it. So, how much plastic do we ingest?

The answer is 184 ng per person per day, or 0.000000184 g.

To help you (and me) to visualise that amount, a grain of salt weighs 60,000 nanograms.

Remember, the WWF says that we ingest 5 g per week, which is what a credit card weighs, when the actual amount is just 0.0000013 g per week. Meaning that it would actually take tens of thousands of years to ingest a credit card's worth of plastic!

Did the WWF, the other NGOs, or the media retract their erroneous claim? Of course not. Good news doesn't sell, and where's the profit in truth?

This is yet more evidence that many so-called "environmental" groups have abandoned the environment in favour of the gravy train — more on that topic later.

The newer, independent review study listed all the sources of microplastics ingested, including fish, molluscs, crustaceans, tap water, bottled water, salt, beer, milk, and the air. Amounts for the individual items listed were extremely low, in the region of 1×10^{-8} to 1×10^{-10} mg/person/day.

Nur Hazimah and Mohamed Nor, Lifetime Accumulation of Microplastic in Children and Adults, Environmental Science & Technology, 55 (8), pp. 5084–5096, 2021

They concluded that amounts are incredibly low compared to inorganic particles.

"Comparing our findings with the intake of other particles, MP mass intake rates are insignificant, as they make up for only 0.001 % of these particles."

This exposes the folly of obsessing over plastic particles. They are 0.001 % of particles we ingest and non-toxic, whereas the other 99.999 % contain proven toxins and carcinogens, meaning substances proven to cause cancer in humans. Ingestion of those other particles, including cancer-causing crystalline silica, is 40 mg per person per day, 200,000 times more than it is for plastic. Anyone focused on the plastic particles and not the real, present danger is exhibiting an irrational fear of plastic.

J. J. Powell et al., Origin and fate of dietary nanoparticles and microparticles in the gastrointestinal tract, Journal of Autoimmunity, 34, pp. 226–233, 2010

There has been an extreme amount of attention on the topic of microplastic in PET bottles. That topic has been studied in huge detail. The particles are safe and they come from the abrasion of the cap made of FDA-approved plastic.

"Microplastic contamination levels in the water were found to increase as the bottle cap is opened and closed repeatedly. The rate of generation of particles with bottle opening and closing cycles (553 ± 202 microplastics/L/cycle) is adequate to account for the total particle density in the water. This clearly demonstrates that the abrasion between the bottle cap and bottleneck is the dominant mechanism for the generation of microplastic contamination detected in bottled water."

T. Singh, Generation of microplastics from the opening and closing of disposable plastic water bottles, Journal of Water & Health, 19.3, pp. 488, 2021

The creation of particles by abrasion can be solved simply by redesigning the screw threads.

A study compared microplastic amounts in one-way PET bottles, returnable PET bottles, glass bottles, and paper-based beverage cartons. They found an amount so low that statistically, they were not more than the control sample, which was ultra-pure filtered deionised water. This is a very important point. So many other studies detect microplastics from, e.g. bottled water, but do not compare the amounts to those found in water that was never in a PET bottle. Dust is everywhere, and methods are now so sensitive that you can detect anything you want anywhere you want. The proposal to avoid PET bottled water makes little sense because particles are there even with no water at all or when a glass bottle or paper-based carton is used instead of PET. They are present in effluent water, ocean water, lake water, river water, canal water, groundwater, and tap water as well because dust is everywhere.

"The average microplastics content was 118 ± 88 particles/l in returnable, but only 14 ± 14 particles/l in single-use plastic bottles. The microplastics content in the beverage cartons was only 11 ± 8 particles/l. Contrary to our assumptions we found high amounts of plastic particles in some of the glass bottled waters (range 0-253 particles/l, mean 50 ± 52 particles/l). A statistically significant difference

from the blank value (14 ± 13) to the investigated packaging types could only be shown comparing to the returnable bottles (p < 0.05)".

D. Schymanski et al., Analysis of microplastics in water by micro-Raman spectroscopy: Release of plastic particles from different packaging into mineral water, Water Research 129, pp. 154–162, 2018

A. A. Koelmans et al., Microplastics in freshwaters and drinking water: Critical review and assessment of data quality, Water Research, 155, pp. 410–422, 2019

In any case, amounts of microplastic from PET bottled water are extremely low, around 0.0000001 % by weight food contact approved polyethylene, and the level of additives found was even lower, around 0.0000000001 %. The media frenzy around these insignificant amounts may well have been fuelled and funded by competitors selling alternative containers made of glass or metal, as there is no rational basis for it.

"Exposure estimations based on the reported microplastic amounts found in mineral water and the assumption of total mass transfer of small molecules like additives and oligomers present in the plastic would not raise a safety concern. Available toxicokinetic data suggests that marginal fraction of the ingested low amount of microplastics can be absorbed, if at all, the conclusion is very likely that the reported amounts present in bottled mineral water do not raise a safety concern for the consumer.

Considering the use of plastic materials in our daily life, occurrence of microplastics in beverages is likely a minor exposure pathway for plastic particles."

F. Welle & R. Franz., Microplastic in bottled natural mineral water — literature review and considerations on exposure and risk assessment, Food Additives & Contaminants: Part A, 35 (12), pp. 2482–2492, 2018

"The estimated daily intake of MPs due to the consumption of bottled water falls within the 4–18 ng kg^{-1} day^{-1} range, meaning that exposure to plastics through bottled water probably represents a negligible risk to human health."

V. Gálvez-Blanca et al., Microplastics and non-natural cellulosic particles in Spanish bottled drinking water, Scientific Reports, 14, 2024

There is no threat according to properly done, peer-reviewed science.

MICROPLASTICS ACCUMULATION

One might wonder what the long-term exposure adds up to over a lifetime. That, too, can be calculated.

We ingest 0.0000013 g per week, and there are around 3600 weeks in 70 years. So, the total lifetime exposure to microplastics by ingestion is less than 0.005 g. The vast majority (~99.7%) of small particles ingested pass right through us. So, we can calculate the total amount not expelled over 70 years as <0.000015 g. We also know that even those tiny amounts not expelled are attacked by our body's defences, degraded, and removed.

T. C. Liebert et al., Subcutaneous Implants of Polypropylene Filaments, Journal of Biomedical Materials Research, 10 (6), pp. 939–951, 1976

Once more, we find that there is no valid reason to be concerned.

Going back to NOAEL, the amount of plastic that can be eaten every day with no effect, which was 50–150 g per day, let us compare that to the actual exposure just mentioned, which is 0.0000002 g. This means that our actual exposure is hundreds of millions of times less than the safe limit.

Anyone genuinely worried about particles should instead focus on the 200,000x greater amount of inorganic particles (with around 1 kg ingested per lifetime) that contain harmful substances, like lead, mercury, and arsenic, plus cancer-causing quartz, than the tiny fraction of non-toxic plastic.

MICROPLASTIC REMOVAL

Have you seen any of the articles where high school students win a prize for inventing a new way to remove microplastics? One such article talks about using ferrofluid to absorb the particles and then remove them with a magnet. I'm not sure who was on the prize committee, but they are clearly not proper scientists. We do not need a new way to remove particles; we have a method that is cheap and works very well. It is called a filter and has been used for centuries. When removing particles from water in a water treatment plant, they coagulate, then filter the water, and that works just fine.

"Results show that on average 89 % of microplastics and 81 % of synthetic fibres (≥63 μm) are retained in water treatment in absence of coagulant. Better final removal efficiency of microplastics (97 %) and synthetic fibres (96 %) was observed in drinking water with coagulation treatment."

A. Velasco et al., Contamination and Removal Efficiency of Microplastics and Synthetic Fibres in a Conventional Drinking Water Treatment Plant, Frontiers in Water, 4, 2022

People are being rewarded for inventing new, but worse, "solutions."

REBRANDING DUST

Thought to have been coined by Professor Richard Thompson in his article "Lost at Sea: Where Is All the Plastic?" published in 2004, the term "microplastic" was in fact first used well over a decade earlier, in 1990, so Thompson is not actually the discoverer after all.

P. G. Ryan & C. L. Moloney, Plastic and other artefacts on South African beaches - temporal trends in abundance and composition, South African Journal of Science, 86, pp. 450–452, 1990

The University of Portsmouth is very proud of Thompson, who has made a career as the supposed father of microplastics. Here's a quote from their website.

> While news articles often describe heavily contaminated locations. It's now clear that plastic and microplastics contaminate shorelines worldwide.
>
> Our work has so clearly shown, that microplastics are present in every sample of beach sand, whether it's in Australia, Asia, Europe, North or South America. We've looked in the deep sea, in Artic ice, in the gut of hundreds of fish from the English Channel, and we've found microplastic contamination everywhere.
>
> Professor Richard Thompson, OBE FRS

Sounds ominous, doesn't it?

Now try replacing the word "microplastic" with the word "dust," and it soon becomes clear just how silly this microplastic hysteria is. We've found dust! Is that worthy of the news? If I call the Editor-in-Chief of *The New York Times* and tell him I have found dust on my keyboard, will it make the front page? Probably not. In fact, they would likely laugh in my face; that is what they should do when people find plastic dust in some new place.

Having said that, here is an actual headline from *National Geographic.*

Microplastics found near Everest´s peak, highest ever detected in the world

E. Napper et al., Reaching New Heights in Plastic Pollution—Preliminary Findings of Microplastics on Mount Everest, One Earth, 3 (5), pp. 621–630, 2020

My response to that was:

"Since when was 'I found dust' news? Dust is everywhere."

Here's another "we found dust" headline.

Newly discovered species found deep in the ocean contains mircoplastic
by Chris Simms

A. J. Jamieson et al., Microplastics and synthetic particles ingested by deep-sea amphipods in six of the deepest marine ecosystems on Earth, The Royal Society, Open Science, 6, 180667, 2019

Why are 99.999 % of particles called "dust" and the other 0.001 % of particles we ingest called "microplastics"? This clever rebranding has enabled some prominent NGOs and some scientists to cash in on our fear.

If you do a Google search for the terms "micrometal," "microwood," "microquartz," and "micropaper," there are no hits (and they show up as spelling mistakes on my computer) because particles of those materials are all just called "dust." The rebranding of one, two hundred thousandth of dust we ingest as "microplastic" has made a mountain out of a molehill and made a fortune for people cashing in on the hysteria.

MICROPLASTICS SCARE STORIES

Part of being a good scientist is to present the data in context so that people can accurately assess the situation. Less ethical scientists show only a part of the picture in order to make their findings seem more important. This latter approach brings fame and funding, so it is easy to see why some people are tempted.

Microplastics in blood

We have all been exposed to headline after headline about microplastics, with no mention of other particles. Why is that? We know that plastic is 0.001 % of the dust we ingest, so why is no one looking for or reporting on the other 99.999 %? Does that sound like good science? I looked and looked for a study that analysed all particles, not just plastic, and finally found one.

This study analysed blood clots from humans and found one particle of poly-ethylene, which we know to be non-toxic, and a vast array of inorganic pigment particles. Phthalocyanine blue pigment is rated as considerably more toxic than plastics or their common additives with a NOAEL of 200 mg/kg/day (OECD). Why do most studies throw away 99 % of particles and only tell you about the plastic ones? Does that seem like good-quality science to you?

"Among twenty-six thrombi, sixteen contained eighty-seven identified particles ranging from 2.1 to 26.0 µm in size. The number of microparticles in each thrombus ranged from one to fifteen with the median reaching five. All the particles found in thrombi were irregularly block-shaped. Totally, twenty- one phthalocyanine particles, one Hostasol-Green particle, and one low-density polyethylene microplastic, which were from synthetic materials, were identified in thrombi. The rest microparticles included iron compounds and metallic oxides."

D. Wu et al., Pigment microparticles and microplastics found in human thrombi based on Raman spectral evidence, Journal of Advanced Research, 49, pp. 141–150, 2023

It would be good to see more professionalism in the future and less "I found dust" or "I found plastic" while omitting to mention or even look for other particles.

nature

Explore content ⌄   About the journal ⌄   Publish with us ⌄   Subscribe

nature > news > article

NEWS | 06 March 2024

Landmark study links microplastics to serious health problems

People who had tiny plastic particles lodged in a key blood vessel were more likely to experience heart attack, stroke or death during a three-year study.

"Landmark study links microplastics to serious health problems!" That was the message we received via the mainstream media following the printing of this headline.

"Presence of microplastics in carotid plaques linked to cardiovascular events"

"In patients with carotid artery disease, the presence of microplastics and nanoplastics (MNPs) in the carotid plaque is associated with an increased risk of death or major cardiovascular events compared with patients in whom MNPs were not detected. This finding supports previous observational data that suggest an increased risk of cardiovascular disease in individuals exposed to plastic-related pollution."

K. Huynh, Presence of microplastics in carotid plaques linked to cardiovascular events, Nature Reviews Cardiology, 21 (5), p. 279, 2024

Based on that, one would have cause for concern. But what does the study really say? When you read the study, the authors specifically say that there's no evidence that the microplastics caused a problem!

As usual, no one actually took the time to read the story before proceeding to spread panic amongst the public.

> *"But Brook, other researchers and the authors themselves caution that this study, published in The New England Journal of Medicine on 6 March, does not show that the tiny pieces caused poor health. Other factors that the researchers did not study, such as socio-economic status, could be driving ill health rather than the plastics themselves, they say."*

R. Marfella et al., Microplastics and Nanoplastics in Atheromas and Cardiovascular Events, The New England Journal of Medicine, 390 (10), 2024

Not only that, but a letter to the editor pointed out that the study was not done properly and may not be credible because of the contamination of the samples.

I wondered whether high particle concentrations can cause cardiac events, and the answer is yes, they can.

Y. Du et al., Air particulate matter and cardiovascular disease: the epidemiological, biomedical and clinical evidence, Journal of Thoracic Disease, 8 (1), pp. 8–19, 2016

But if there are 200,000 other inorganic particles per one plastic particle, why on Earth would any sane person assume that the plastic particle is to blame?! The answer is that they wouldn't because there is no evidence to support that hypothesis.

Such hysterical stories often say this is "linked" to that or "associated" with this, but that is meaningless. Two events occurring together do not mean that one caused the other. If I go for a walk and it's sunny, do my neighbours assume I made it sunny? I hope not because that would be really silly.

All good scientists know, as Brook pointed out, that correlation does not imply causation.

There is a famous cartoon showing that shark attacks and ice cream sales are correlated, and a layperson might be tempted to think that one must cause the other. In fact, they are correlated because both happen when people go to the beach when the weather is nice. There are more shark attacks simply because there are more people in the water when the sun is out. It has nothing to do with eating ice cream. Therefore, we must be wary when we are told that A is "linked" to B. Often, they are linked in some way, but one is not the cause of the other. Scientists remind us that correlation does not mean causation.

EAT ICE CREAM → GET ATTACKED BY SHARK

Microplastics in the brain

There have been several media headlines about plastic particles moving around the body. I was surprised too. The stories portray this as some new and alarming discovery that is specific to plastic particles. Is that the case? I am not a biologist, so I had to check the science to find out.

I was amazed to learn that particles entering the body and moving is called "translocation" and has been studied for almost 200 years. So, it is not new, and it is not specific to plastics either because they had not been invented in 1844. Quite the contrary, translocation has been reported for all kinds of particles.

E.F.G. Herbst, In: Das Lymphgefasssystem und seine Verrichtungen, (Eds. Vandenhoek and Ruprecht), Gottingen, pp. 333–337, 1844

More recently, various studies have continued to show all kinds of particles in the body are moving around.

"These results demonstrate effective translocation of ultrafine elemental carbon particles to the liver by 1 d after inhalation exposure."

G. Oberdörster et al., Extrapulmonary translocation of ultrafine carbon particles following whole-body inhalation exposure of rats, Journal of Toxicology & the Environment Health A., 65 (20), pp. 1531–43, 2002

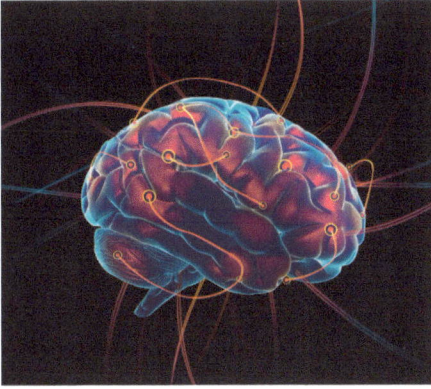

For context, we can check historical studies. For example, the following study also detected the movement of particles into the brain.

"There was a significant and persistent increase in added ^{13}C in the olfactory bulb of 0.35 µg/g on day 1, which increased to 0.43 µg/g by day 7. Day 1 ^{13}C concentrations of cerebrum and cerebellum were also significantly increased but the increase was inconsistent, significant only on one additional day of the postexposure period, possibly reflecting translocation across the blood–brain barrier in certain brain regions."

G. Oberdörster et al., Translocation of inhaled ultrafine particles to the brain, Inhalation Toxicology, 16, pp. 437–445, 2004

What about the scare stories reporting for the first time that plastic particles can enter the brain? Sounds scary, but instead of reacting to the headline, it is wise to look a little deeper. Is this new information worthy of an immediate response, or is it out of context?

"Micro- and Nanoplastics Breach the Blood–Brain Barrier (BBB): Biomolecular Corona's Role Revealed"

V. Kopatz et al., Micro- and Nanoplastics Breach the Blood–Brain Barrier (BBB): Biomolecular Corona's Role Revealed, Nanomaterials, 13, 1404, 2023

In the study, they force-fed mice with an insanely high concentration of lab-made polystyrene particles unlike any particles found in the environment. The unrealistic dose meant that the body's defence system was overwhelmed, so the particles reached the brain. However, the study tells us nothing about actual exposure conditions and is pretty much meaningless.

The Oberdörster group continued to investigate translocation (movement) of particles in the body. They cited a study as far back as 2002, over two decades ago, showing that polystyrene was one such type of nanoparticle among several others, including gold, iridium, and carbon. This shows that the "discovery" of synthetic polystyrene nanoparticles crossing into the brain of rodents is not new at all, but is, in fact, over 20 years old.

G. Oberdörster et al., Nanotoxicology: An Emerging Discipline Evolving from Studies of Ultrafine Particles, Environmental Health Perspectives?, 113 (7), July 2005

The paper on synthetic polystyrene nanoparticles in hamsters was very informative. Nemmar et al. showed that the surface charge of the synthetic polystyrene particles determined their behaviour in the body. This is a key point because laboratory-synthesised polystyrene particles of the type used in the new 2023 study are unlike the kind of polystyrene found in the environment. The lab particles have a charge intentionally added, which makes them interact much more than real uncharged polystyrene particles do. This charge effect has been confirmed by other researchers.

A. Nemmar et al., Ultrafine Particles Affect Experimental Thrombosis in an In Vivo Hamster Model, American Journal of Respiratory and Critical Care Medicine, 66, pp. 998–1004, 2002

S. Wieland et al., Nominally identical microplastic models differ greatly in their particle-cell interactions, Nature Communications, 15 (922), 2024

This reinforces the point that studies on lab-made polystyrene are not relevant for understanding what really happens in the environment. For that matter, scientists have also noted that polystyrene itself is the wrong type of plastic to use because the plastics in the environment are not polystyrene but rather dominated by polyethylene (PE), polypropylene (PP), and polyethylene terephthalate (PET).

K. Tanaka and H. Takada, Microplastic fragments and microbeads in digestive tracts of planktivorous fish from urban coastal waters, Scientific Reports 6(1):34351, 2016

We can now see that the dramatic headlines about microplastics in the brain were unfounded for multiple reasons.

- The study was invalid due to unrealistically high particle concentrations.
- They used the wrong plastics — a type no one is exposed to in the real world.

- The existence of particles in the body is not news because it has been known for decades.
- The movement of particles around the body, including the brain, is 20-year-old news.
- The same effect happens with all kinds of particles.

Here are some studies spanning many years about other common particles doing the exact same thing that plastics are found to do.

Manganese oxide nanoparticles

"We conclude that the olfactory neuronal pathway is efficient for translocating inhaled Mn oxide as solid UFPs to the central nervous system and that this can result in inflammatory changes. We suggest that despite differences between human and rodent olfactory systems, this pathway is relevant in humans."

A. Elder et al., Translocation of Inhaled Ultrafine Manganese Oxide Particles to the Central Nervous System, Environmental Health Perspectives, 114 (8), 2006

Carbon black nanoparticles

This one is really pertinent because carbon black makes car tyres black and is abraded into the atmosphere in high amounts.

"Higher levels of black carbon predicted decreased cognitive function across assessments of verbal and nonverbal intelligence and memory constructs."

S. F. Suglia et al., Association of Black Carbon with Cognition among Children in a Prospective Birth Cohort Study, American Journal of Epidemiology, 167, pp. 280–286, 2008

Zinc oxide nanoparticles

Zinc oxide is used in physical sunscreens, so we are exposed to it.

> *"Our results suggest that acute exposure to ZnONP induces oxidative stress, microglia activation, and tau protein expression in the brain, leading to neurotoxicity."*

H.C. Chuang et al., Acute Effects of Pulmonary Exposure to Zinc Oxide Nanoparticles on the Brain in vivo, Aerosol and Air Quality Research, 20, pp. 1651–1664, 2020

Iron soot nanoparticles

> *"Our findings visually demonstrate that inhaled ultrafine iron-soot reached the brain via the olfactory nerves and was associated with indicators of neural inflammation."*

L. E. Hopkins et al., Repeated Iron-Soot Exposure and Nose-to-Brain Transport of Inhaled Ultrafine Particles, Toxicologic Pathology, 46 (1), pp. 75–84, 2018

Titanium dioxide nanoparticles

This is the most common white pigment found in paper, plastics, and physical sunscreens.

> *"…in the rat, spherical, small TiO_2-NPs significantly increased the BBB permeability and entered the brain. TiO_2-NPs were accumulated in the brain, but no obvious pathological anomaly was observed in the cerebral cortex and hippocampus."*

X. Liu et al., Size- and shape-dependent effects of titanium dioxide nanoparticles on the permeabilization of the blood-brain barrier, Journal of Materials Chemistry B, 48, 2017

While looking into the science on this topic, I also learned that the ability of nanoparticles to cross into the brain is exploited by scientists — they employ such particles to deliver drugs targeted to the brain. There are quite a few studies on the subject.

While it may be unsettling to think of particles inside our bodies, it is important to note that all particles do it, and our bodies are used to dealing with it. We have developed immune systems that can envelop particles for removal or attack and destroy them.
As we have seen, particulate pollution is a real problem. It is appropriate to study it and evaluate the risks. However, it is not appropriate to obsess over plastics, which are just 0.001 % of the particles we ingest.

Nur Hazimah and Mohamed Nor, Lifetime Accumulation of Microplastic in Children and Adults, Environmental Science & Technology, 55 (8), pp. 5084–5096, 2021

It is also not meaningful to scare the public over particles they will never encounter in the real world. Why scare the public with 20-year-old news when we should focus on real and present dangers?

Microplastic in the placenta

This topic is like the case of particles in the blood and in the brain. It is not news and occurs for all kinds of particles, including carbon black pigment found in car tyres.

H. Bové et al., Ambient black carbon particles reach the fetal side of human placenta, Nature Communications, 10, 3866, 2019

The same has been reported long ago for silica (which sand is made of)

and titanium dioxide, which is a very common white pigment used in sunscreen.

K. Yamashita et al., Silica and titanium dioxide nanoparticles cause pregnancy complications in mice, Nature Nanotechnology, 6, pp. 321–328, 2011

Silver, silica, carbon, alumina, cerium oxide, diesel exhaust, quantum dots, platinum, titanium dioxide, gold, iron oxide, polystyrene, fullerenes, zinc oxide, zirconium oxide, and carbon nanotubes have all been reported in the placenta.

T. Buerki-Thurnherr et al., Knocking at the door of the unborn child: engineered nanoparticles at the human placental barrier, Swiss Medicinal Weekly, 142, 2012

E. Bongaerts et al., Translocation of (ultra) fine particles and nanoparticles across the placenta; a systematic review on the evidence of in vitro, ex vivo, and in vivo studies, Particle & Fibre Toxicology, 17 (56), 142235, 2020

Dust gets everywhere so that is not so surprising. It is not responsible reporting to act as though this is something new, specific to plastic particles, and dangerous when it is not any of those things.

TOXINS & MICROPLASTICS

Yet another claim is that microplastics release toxic chemicals, but as we saw earlier in the chapter, plastics and typical additives are non-toxic. So, what do these claims refer to?

One common idea is that fish eat microplastics and are thereby exposed to toxic chemicals. However, closer examination reveals that the chemicals are actually from the ocean water and not the plastic. Such chemicals are absorbed by plastic because "like dissolves like"; this saying refers to the fact that fatty substances (hydrophobic is the scientific term) prefer to be inside the fatty (hydrophobic) polymers, so they leave the sea water in which they are poorly soluble and concentrate inside the plastic instead.

Once more, the NGOs have distorted reality to paint plastics as the villain. NGOs claim that the plastic acts as a "vector" for transporting toxins, but what do studies on toxins and microplastics say? They show that toxic chemicals in the ocean are absorbed by the plastic and are thus removed from the water. The result is that the marine organisms are protected because the poison is now inside the plastic microparticles (MP) and is no longer in the water. That's the opposite of what the NGOs say.

"Both test species actively ingested the MP particles. However, the presence of MP never increased the bioaccumulation of neither model chemicals, nor their toxicity to the exposed organisms. Bioaccumulation was a linear function of waterborne chemical disregarding the level of MP. Toxicity, assessed by the threshold (EC$_{10}$) and median (EC$_{50}$) effect levels, was either independent of the level of MP or even in some instances significantly decreased in the presence of MPs. These consistent results challenge the assumption that MP act as vectors of hydrophobic chemicals to planktonic marine organisms."

R. Beiras et al., Polyethylene microplastics do not increase bioaccumulation or toxicity of nonylphenol and 4-MBC to marine zooplankton, Science of the Total Environment, 629, pp. 1–9, 2019

This next study came to the same conclusion.

"The addition of microplastics to synthetic water significantly reduced the toxicity of bifenthrin (apparent LC50 = 1.3 µg/L), most likely because sorption of bifenthrin to microplastics reduced its bioavailability to the exposed larvae. A sorption capacity experiment showed that N92% of bifenthrin was sorbed to microplastics."

The plastic removed 92% of the toxin. The workers made another important point, which is that in the real world, there are so many other types of organic particles around (leaves, sticks, etc.) that the effects of plastic are negligible anyway.

"Strikingly, the addition of microplastics to river water did not mitigate bifenthrin toxicity (apparent LC50 = 1.4 µg/L), most likely due to greater interaction of bifenthrin with organic carbon than with microplastics."

S. Ziajahromi et al., Effects of polyethylene microplastics on the acute toxicity of a synthetic pyrethroid to midge larvae (Chironomus tepperi) in synthetic and river water, Science of the Total Environment, 671, pp. 971–975, 2019

This highlights how misleading it is to talk about plastic particles while forgetting how insignificant their concentration is in the wider picture.

This next study also found that while plastics absorb toxins and provide a protective effect in the lab, in real-world situations, interaction with natural particles is the major factor.

"Low microplastic concentrations loaded with phenanthrene or anthracene induced a less pronounced response in the sediment communities compared to the same total amount of phenanthrene or anthracene alone."

"Due to high ambient concentrations of organic pollutants and their sorption to natural particles, the transported amounts of two PAHs (anthracene and phenanthrene) did not add substantial quantities to background environmental levels in the sediment."

J. Kleinteich et al., Microplastics Reduce Short-Term Effects of Environmental Contaminants. Part II: Polyethylene Particles Decrease the Effect of Polycyclic Aromatic Hydrocarbons on Microorganisms, International Journal of Environmental Research and Public Health, 15 (287), 2018

Probably the most detailed examination of this topic was a review article, which pointed out that all other studies assume that 100% of chemicals would migrate out of microplastic after ingestion; in reality, that does not occur because there is not enough time during digestion. They found that actual exposure levels are vastly lower, so low as to be "small to negligible."

"Previous risk assessments that evaluate the role of MPs as chemical vectors in humans have so far assumed worst case scenarios in their calculations, with 100% instantaneous leaching of chemicals. In the present study, we performed a probabilistic assessment to evaluate the actual chemical exposure via MPs in relation to dietary and inhalation intake of compounds using the simulated MP intake rates and also accounting for the full variability of the MP continuum. Our methodology also includes quantifying the actual percentage change in the body tissue concentrations with the added chemicals from MP intake. We conclude that the contribution of the MPs to chemical intake is small to negligible for the four representative chemicals investigated in this study…"

Nur Hazimah Mohamed Nor, Lifetime Accumulation of Microplastic in Children and Adults, Environmental Science & Technology, 55 (8), pp. 5084–5096, 2021

An emphasis in this book is zooming out from the plastics-only discussion to get a more balanced viewpoint by comparing plastics with our other material options. Here is a study that compares the chemicals coming out of glass bottles into our drinking water to what happens when we choose PET bottles instead.

"Many more elements leach from glass than from PET bottles. Comparing the same water sold in PET bottles to results for water sold in glass bottles Ce, Pb, Al and Zr are the 4 elements that leach most from glass, but Ti, Th, La, Pr, Fe, Zn, Nd, Sn, Cr, Tb, Er, Gd, Bi, Sm, Y, Lu, Yb, Tm, Nb and Cu are all significantly enriched in the glass bottles when compared to the same water sold in PET bottles."

C. Reimann et al., Bottled drinking water: Water contamination from bottle materials (glass, hard PET, soft PET), the influence of colour and acidification, Applied Geochemistry, 25, pp. 1030–1046, 2010

In case your chemistry is rusty, they are saying that the metals coming out of glass bottles and contaminating water are far worse than the plastic coming from PET bottles. Metals found are cerium, lead, aluminium, and zirconium, with many other heavy and transition metals leaching from glass as well. When is the last time you read an article in the newspaper or online that mentioned that? Perhaps the glass industry has much better lobbyists to control what we see?

While we are on the topic of perspective, it is worth saying something about the concept of "detection." We see stories that microplastics were "detected" here or that a toxic chemical was "detected" there. Scientists love to detect things, and the machines they use have grown ever more sensitive. In fact, they are now so sensitive that you could probably detect almost anything you wanted to, almost anywhere. That may sound like an exaggeration, but let me give you an example.

I read a study in which they had detected some kind of chemical coming from microplastics and the concentration was about 1 ng/L (one nanogram per litre). Even as a PhD chemist, I had a hard time visualising how much that really is, so I ran a calculation. Turns out it is such an incredibly low amount that it is almost ridiculous.

A nanogram per litre is one millionth of one part per million. Imagine taking an object, cutting it into a million pieces, taking one of those pieces and cutting it into a million pieces, then selecting just one of those pieces. Concentrations that low are not worth scaring the public over, but that does not stop some scientists and NGOs from doing just that.

All of this chemistry talk may seem bewildering, so I came up with another analogy. The population of the entire planet is around 8 billion people, so what is a millionth of a millionth of that?

A millionth of 8 billion people is 8000 people.

A millionth of that is 0.008 people.

An average person weighs 70 kg. 0.008 of 70 kg is about half of one kilogram (about 1 lb).

A human hand weighs the same amount.

So, starting with the entire population of the world, a millionth of a millionth is the same weight as one human hand.

These are the vanishingly small amounts that we can now detect. Scientists really should be more responsible before proclaiming that they "detected" a substance. We need not just data but a responsible amount of perspective to go with the data.

BAD SCIENCE

In *The Plastics Paradox* book and the website of the same name, I called out the appallingly bad science in the microplastics field. I analysed study after study, finding errors so serious as to instantly invalidate the study. Perhaps people thought I was being too harsh. However, in the years since, other scientists, including Lenz et al., Gouin et al., and Koelmans et al., have made the same observations; they called out the fact that most studies use a kind of special plastic particle that is not even present in the environment then they use a million times too much of it. Some studies even soak the plastic particles in poison so that they can claim that the plastic is poisonous. Two detailed reviews agree with my assessment and find that 85–92% of microplastics studies are flawed for the very reasons I have been stating for years.

What concentration of microplastic should scientists use to create a valid, realistic study? Lenz et al. provided an important contribution on that subject.

"Microplastic research is an emerging field, and there is a lot of misunderstanding and in some cases over-reaction or misinterpretation of results from MP science in the public. We therefore strongly suggest that future studies of MP impact on marine ecosystems should also include concentrations that have been documented in the environment to yield more realistic estimates of sublethal effects."

"Experimental exposure concentrations tend to be between two to seven orders-of-magnitude higher than environmental levels."

R. Lenz, K. Enders, and T. G. Nielsen, Microplastic exposure studies should be environmentally realistic, Proceedings of the National Academy of Sciences, 113 (29), E4121–E4122, 2016

They point out that studies use up to ten million times too much plastic compared to the amount that would accurately represent the amount that is really in the environment. Lenz implored other scientists to do proper science at proper concentrations. Any competent toxicologist will tell you that using such high concentrations means that the study is invalid. I would call it junk science, and it is one of the crucial mistakes that invalidates studies on this topic.

Sometimes, the scientists make other errors. One relatively common error is to detect particles and then claim they are microplastics without ever checking to make sure that they are in fact made of plastic. This is science so poor that words almost fail me, and yet, this theme reoccurs. This study claims to have found incredibly high numbers of plastic particles in fruit and vegetables in shops.

"The higher median (IQR) level of MPs in fruit and vegetable samples was 223,000 (52,600–307,750) and 97,800 (72,175–130,500), respectively. In particular, apples were the most contaminated fruit samples, while carrot was the most contaminated vegetable. Conversely, the lower median (IQR) level was observed in lettuce samples 52,050 (26,375–75,425)."

G. O. Conti et al., Micro- and nano-plastics in edible fruit and vegetables. The first diet risks assessment for the general population, Environmental Research, 187, 2020

Such stories go viral, but no one seems to read them to make sure that the science is sound. They dissolved the fruit and vegetables in concentrated acid and then incorrectly assumed that they must be made of plastic. Perhaps these scientists should have paid more attention in school.

There are too many examples of this bad science to recount them all, but here is what two reviews found. This first one showed that only around 10% of microplastic studies are done on the right kinds of plastic, namely the PE, PP, PVC, and PET that is present in the environment.

">80% of studies are identified as not reliable"

"…few studies provide information that support that the particles tested are representative of NMPs found in the environment, or that the concentrations tested are representative of environmentally relevant exposure scenarios."

T. Gouin et al., Screening and prioritization of nano- and microplastic particle toxicity studies for evaluating human health risks — development and application of a toxicity study assessment tool, Microplastics & Nanoplastics, 2 (2), 2022

A simple analogy might help to highlight why it is so important to do testing on the right kinds of plastic. If you wanted to know whether kittens are dangerous, would you study kittens or lions?

"Microplastics are frequently present in freshwaters and drinking water, and number concentrations spanned ten orders of magnitude (1×10^2 to 108 #/ m^3) across individual samples and water types. However, only four out of 50 studies received positive scores for all proposed quality criteria, implying there is a significant need to improve quality assurance of microplastic sampling and analysis in water samples."

A. A. Koelmans et al., Microplastics in freshwaters and drinking water: Critical review and assessment of data quality, Water Research, 155, pp. 410–422, 2019

They reported that just 8% of studies were reasonable quality and the other 92% were lacking and thus unreliable. As shown in other studies, particles were found in lake, river, ground, tap, and bottled water. As expected, dust is everywhere and the current paranoia around bottled water and microplastic is unwarranted.

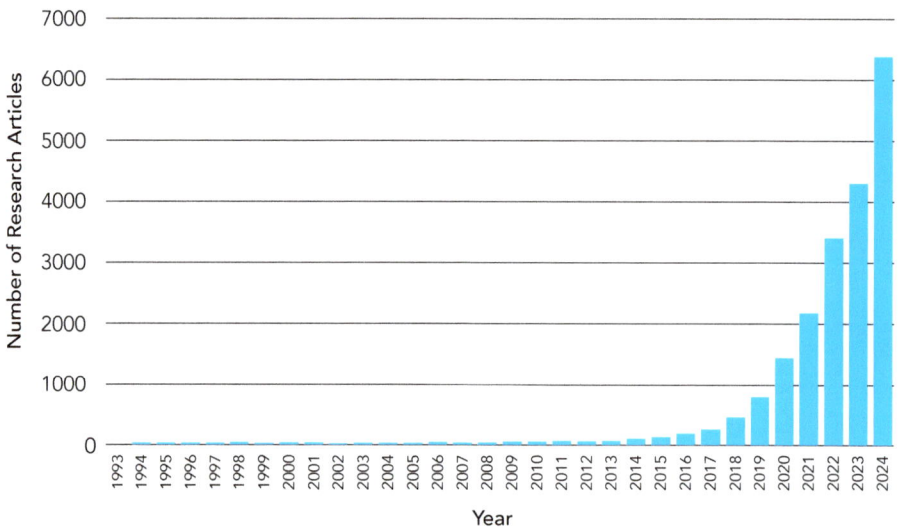

We have witnessed accelerating growth in the number of microplastic studies per year. Many would argue that is a good thing. After all, should we not study anything that may present a danger to us? Some argue that more knowledge can only be helpful. Sounds reasonable, doesn't it?

It does sound reasonable until we look at the cost of all those studies that are paid for with our taxes. Unlimited anything sounds great until the cost is factored in. Plus, what quality is this information that we are paying for?

6400 studies a year on microplastics at around $30,000 per study (an estimate from a professor) means almost $200 million a year of our tax money. That's a lot.

We just saw that two reviews concluded that up to 90% of the studies are flawed / not valid. So, we are wasting around $180 million a year on bad science.

Not only that, but we already have the studies we need to reveal the amount we're exposed to (extremely low) and the level of threat (non-toxic like clay and cellulose).

"…the experimental design of most studies does not allow distinguishing plastic-specific effects from those caused by any other particles, such as clay and cellulose, which are ubiquitously present in the environment. We suggest that microplastic effects reported in recent ecotoxicological studies are similar to those induced by the natural particles."

M. Ogonowskia, Z. Gerdesa & E. Gorokhova, What we know and what we think we know about microplastic effects — A critical perspective, Current Opinion in Environmental Science & Health, 1, pp. 41–46, 2018

Studies conclude plastics like PE, PP, PVC, and PET are not toxic — no matter whether they are particles or fibres.

"This work for the first time investigated and compared the intestinal uptake and cytotoxicity of microplastic particles of the commonly produced materials PE, PP, PVC and PET in vitro."

"None of the particles triggered acute toxic effects, regardless of their shape and material."

"Only excessively high concentrations far beyond realistic dietary exposure of consumers induce cytotoxic effects."

V. Stock et al., Uptake and cellular effects of PE, PP, PET and PVC microplastic particles, Toxicology in Vitro, 70, 105021, 2021

"The results revealed no adverse effects of secondary microplastics (PP and PS) as determined by clinical signs, body weights, or organ weights and no gross pathological findings in any of the treatment groups. This study will provide basic data for sub-chronic and chronic repeated dose toxicity of microplastics."

J. Sik-Kim, Acute toxicity evaluation of polypropylene and polystyrene microplastics in Sprague Dawley (SD) rats after oral administration, Journal of Pharmacological and Toxicological Methods, 105, 106813, 2020

How about we stop wasting money on bad science done on a topic that's already been covered? I know some people will reply that science never stops, and we might discover some new threat, but that argument does not hold water. If plastic dust was especially toxic, the last 50 years of studies would have already said so, but they didn't. Repeating the same studies makes as much sense as paying someone to drop an apple all day every day just to check that Isaac Newton was correct and that gravity exists.

MICROPLASTIC DEGRADATION

The general perception is that plastics never really degrade — instead, they fragment into smaller and smaller pieces and then stop. This, of course, completely defies logic and our own experience with other materials. Do leaves crumble into pieces and then stop degrading at a certain size? Do cars start rusting and then magically stop? No, they don't, and you would be called a fool if you declared they did, and yet, that is precisely what NGOs claim about plastics.

Scientists have shown that microplastics continue to degrade until they form water and carbon dioxide, which is what all other organic materials do, meaning they degrade to the same final products as paper, apples, leaves, and trees. All organic matter (PE, PP, PET, apples, leaves, cotton) is based on the same elements, including carbon, hydrogen, and oxygen. This common chemistry is what makes them degrade similarly.

"Microplastic debris in the environment degrades mechanically, chemically, and biologically."

"Microplastics degrade through the same processes that break down macro-plastic debris items, albeit more quickly because of their higher surface to volume ratio."

"Carbon dioxide, H_2O, and CH_4 are produced in this final step known as mineralization."

J.C. Prata in T. Rocha-Santos, M. Costa, C. Mouneyrac (eds), Handbook of Microplastics in the Environment, Springer Switzerland, pp. 531–542, 2022

A. Delre et al., Plastic photodegradation under simulated marine conditions, Marine Pollution Bulletin, 187, 2023

What about microplastics in the ocean? Do they degrade too? This study looked at the most common plastics, including LLDPE, PP, EPS (polystyrene foam), PET, PVC, PA, and PCL.

"Using real world data, we reveal that plastic surfaces can degrade at a rate of up to 469.73 µm per year, 12 times greater than previous estimates."

C. Maddison et al., An advanced analytical approach to assess the long-term degradation of microplastics in the marine environment, Materials Degradation, 7 (59), 2023

Not only do plastics degrade in the oceans, but they do so over 10 times more rapidly than originally assumed. Rather than endlessly accumulating, as is claimed, amounts found are low, and they are removed by degradation, just the same as other materials.

THE MISSING FACTOR

There is one aspect in the discussion around microplastics that I have never seen mentioned. Imagine that plastic was replaced because of concerns over microplastics — would that be a positive move? Well, we know it takes 3–4 times more paper, metal, wood or glass to replace plastic and that those materials also degrade to form particles. In the case of wood, those particles are known to cause cancer. Copper dust is highly toxic too. So, replacing plastic would increase the quantities of particles we are exposed to, and the average toxicity of those particles. Does that sound wise? People are so eager to be against plastic that they almost never stop to consider the consequences of moving to alternatives.

SUMMARY

Fear is not rational, and it is not easy to convince someone not to be frightened. We have thousands of phobias, from arachnophobia (fear of spiders) to xenophobia (fear of foreigners). We can add plastiphobia to that list. People have been intentionally misled into fearing plastics when decades of science show that there is no rational reason for that fear. Hopefully, those of you who have read this far have been reassured by the huge amounts of peer-reviewed evidence. This is not some new, previously unidentified problem. On the contrary, we have 50 years of studies on plastic particles — amounts are low, and they are non-toxic. We are only concerned because NGOs cleverly rebranded plastic dust to make it sound scary and were helped by the media, who abandoned the truth long ago.

One crucial factor when evaluating risk is perspective. If we cannot prioritise large, genuine threats over insignificant or imaginary ones, then we will end up paralysed, hiding under a blanket and afraid to venture outside lest the sky fall on our heads. In the name of perspective, here is a breakdown of what people really die from. It is not plastic, microplastic, or parts per million of chemicals. Anyone who truly desires a safer, healthier life can glance at this list to see what needs to be done.

What do people die from? Causes of death globally in 2019

The size of the entire visualization represents the total number of deaths in 2019: 55 million.
Each rectangle within it is proportional to the share of deaths due to a particular cause.

74% dies from noncommunicable diseases   **14% dies from infectious diseases**

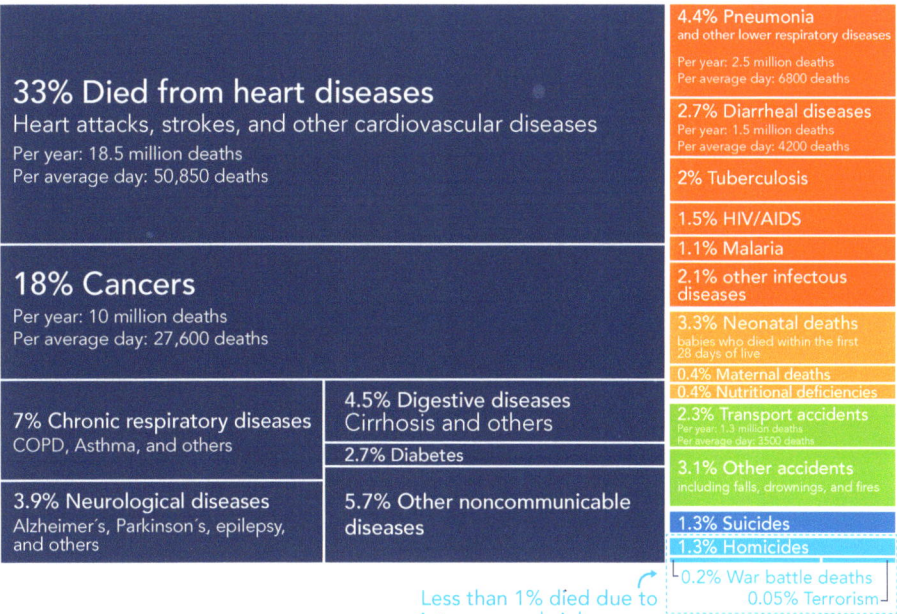

33% Died from heart diseases
Heart attacks, strokes, and other cardiovascular diseases
Per year: 18.5 million deaths
Per average day: 50,850 deaths

18% Cancers
Per year: 10 million deaths
Per average day: 27,600 deaths

7% Chronic respiratory diseases
COPD, Asthma, and others

4.5% Digestive diseases
Cirrhosis and others

2.7% Diabetes

3.9% Neurological diseases
Alzheimer's, Parkinson's, epilepsy, and others

5.7% Other noncommunicable diseases

4.4% Pneumonia
and other lower respiratory diseases
Per year: 2.5 million deaths
Per average day: 6800 deaths

2.7% Diarrheal diseases
Per year: 1.5 million deaths
Per average day: 4200 deaths

2% Tuberculosis

1.5% HIV/AIDS

1.1% Malaria

2.1% other infectous diseases

3.3% Neonatal deaths
babies who died within the first 28 days of live

0.4% Maternal deaths
0.4% Nutritional deficiencies

2.3% Transport accidents
Per year: 1.3 million deaths
Per average day: 3500 deaths

3.1% Other accidents
including falls, drownings, and fires

1.3% Suicides
1.3% Homicides

0.2% War battle deaths
0.05% Terrorism

Less than 1% died due to interpersonal violence

Data source: IHME Global Burden of Disease and Global Terrorism Database
OurWorldinData.org – Research and data to make progress against the world's largest problems.
Licensed under CC-BY by the author Max Roser

People are terrible at accurately gauging risk, which is why numbers help us to focus on what matters. In 2019, over 60,000 people died from snakebites, which equates to around 3 million years of life lost, whereas recorded mortalities from microplastics were zero. This example emphasises the importance of real risk over imaginary risk.

N. L. S. Roberts et al., Global mortality of snakebite envenoming between 1990 and 2019, Nature Communications, 13, 6160, 2022

Just recently, someone said that they hope I am not offended that they keep asking questions about microplastics after I had provided several links to the science. He clearly had not looked because his questions were already addressed in the links provided. My reply was:

"I am not at all offended. Anyone concerned can look at the science provided, see the facts and be reassured. Or they can avoid looking and continue to be worried needlessly. It's up to each person to decide."

We need to recognise that there are many types of people. Some cannot be reached with facts, and others enjoy being scared for no reason. These are the people who pay money to see horror movies and be scared senseless. Each to their own.

Sensible actions to improve one's health would be to go easy on the pizza, take a walk every day, don't smoke, and don't drink too much. These simple, easy steps will have far more benefit than anything else. Of course, fretting over trivial things like straws or plastic dust is less work than addressing real issues.

When it comes to policy, there is no evidence to suggest that any policy changes are needed. However, there certainly is an urgent need for other kinds of action.

We need to expose and shut down NGOs that frighten us and our children with lies. We should impose heavy fines on journalists and media outlets that mislead us. We should impose heavy fines on academics who conduct junk science experiments with a million times too much plastic.

Let's create a better future based on truth and wisdom instead.

~90% of the science on microplastics is worthless, and the studies relayed to us are only the scary ones because that's how the media and NGOs make money. When we really read the studies and find the reliable ones with proper scientific methods, we see 50 years of data and no credible evidence of harm. The FDA agrees.

MAKING RESPONSIBLE CHOICES WITH LIFE CYCLE ANALYSIS

MAKING RESPONSIBLE CHOICES WITH LIFE CYCLE ANALYSIS

A wise man once said:

"There are no solutions. There are only trade-offs."

Thomas Sowell, A Conflict of Visions: Ideological Origins of Political Struggles,
Basic Books, New York, NY, USA, 2007

The same holds true for materials — there is no perfect material. Would you make a teapot out of chocolate? I hope not.

The best that a wise person can do is to select the option proven to cause the least impact, as long as it is fit for purpose. But how can we know which one is best? Scientists have an answer for that, and it is called "life cycle analysis," or LCA for short.

LIFE CYCLE ANALYSIS OF PLASTIC COMPARED TO ALTERNATIVES

Every action we take has an associated impact, and all materials create an impact, too. Smart people who care know that the best path is to minimise that impact by selecting the alternative that does the least harm. Life cycle analysis is the only proven method for comparing impact. It is accepted worldwide by companies, governments, and NGOs. Not only that, but it has been honed over decades. It is standardised, and the information is drawn from established, credible, and shared databases. Once the LCA is ready, it has to be independently checked to ensure there is no funny business going on.

As shown in the diagram, the LCA methodology is to consider every step in the manufacture, use, and disposal of a product, which could be anything, such as a car, a washing machine, or a coffee cup. By adding up the environmental impact of every stage, we can work out which option has the least impact and pick that one.

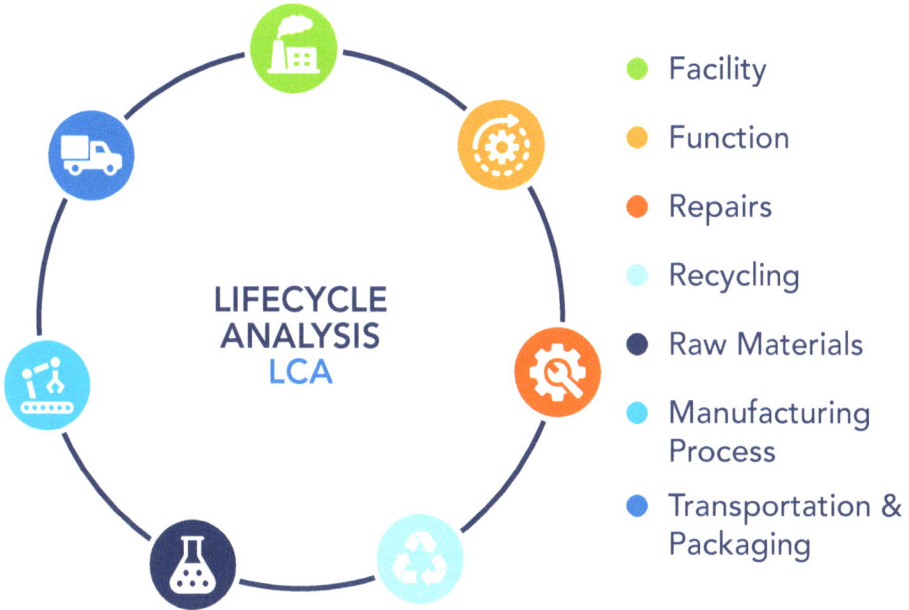

LIFECYCLE
ANALYSIS
LCA

- Facility
- Function
- Repairs
- Recycling
- Raw Materials
- Manufacturing Process
- Transportation & Packaging

• Carbon Dioxide   • Energy Used   • Water Used
• Acid Rain   • Toxicity   • Eutrophication   • Waste

So, what do the largest reviews comparing life cycle studies on plastics with those on alternative materials say?

"This review analysed 53 peer-reviewed studies published in the time range 2019–2023, aiming at understanding the state of the art in LCA about the environmental impacts of packaging by focusing on the comparison between plastics and alternative materials. The literature showed that consumer perceptions often differ from LCA findings and revealed that, frequently, conventional plastics are not the least environmentally friendly choice."

"With regard to the materials comparison, the review led to the conclusion that, despite the common sense, plastic is not the most impacting option. Accordingly, the compared materials do not generally appear to be friendlier than plastics from the environmental perspective."

D. Dolci et al., How does plastic compare with alternative materials in the packaging sector? A systematic review of LCA studies, Waste Management & Research, pp. 1–19, 2024

The scientists are politely saying that the public believes the opposite of what is actually true and that being against plastics means increasing impact, not the reverse. Note the tremendous weight of their comprehensive study, which reviewed 53 separate LCAs.

Here is another huge review of life cycle studies, where they looked at 16 different applications: shopping bags, wet pet food packaging, soft drink containers, fresh meat packaging, industrial drums, soap containers, milk containers, water cups, municipal sewer pipes, residential water pipes, building insulation, furniture, hybrid fuel tanks, BEV battery enclosures, carpets, and t-shirts.

"We assess 16 applications where plastics are used across five key sectors: packaging, building and construction, automotive, textiles, and consumer durables.

These sectors account for about 90% of the global plastic volume.

Our results show that in 15 of the 16 applications, a plastic product incurs fewer GHG emissions than their alternatives.

In these applications, plastic products release 10% to 90% fewer emissions across the product life cycle."

F. Meng et al., Replacing Plastics with Alternatives Is Worse for Greenhouse Gas Emissions in Most Cases, Environmental Science & Technology, January, 2022

They concluded that in 15 out of 16 cases, the plastic option caused the least impact. That's 93% of the time that choosing an alternative to plastic makes matters worse.

"These results demonstrate that care must be taken when formulating policies or interventions to reduce plastic use so that we do not inadvertently drive a shift to nonplastic alternatives with higher GHG emissions. For most plastic products, increasing the efficiency of plastic use, extending the lifetime, boosting recycling rates, and improving waste collection would be more effective for reducing emissions."

This huge review covering 73 life cycle reports was mentioned at the beginning of the book, but the finding is so important that it bears repeating.

"Several studies have shown many materials used as alternatives to plastic in packaging, such as cotton, glass, metal or bioplastics, to have significantly higher CO_2 impact or water usage compared to plastic packaging. On average over current food packaging, replacing plastic packaging with alternatives, would increase the weight of the packaging by 3.6 times, the energy use by 2.2 times, and the carbon dioxide emissions by 2.7 times"

N. Voulvoulis et al., Examining Material Evidence — The Carbon Footprint, Centre for Environmental Policy, Imperial College London & Veolia UK, ACC, 2019

There certainly are alternatives to plastic, but they almost always make matters worse, not better. Now, let us look in more detail at some specific, high-profile examples.

PET bottles

To highlight the folly of moving from PET bottles to alternatives, here is a quote from that last review.

"When considering the production and manufacturing of the main alternatives to plastic for a 500ml bottle, other packaging types (fibre, glass, steel and aluminium) emit more greenhouse gases than plastic bottles, with glass bottles being the highest emitter overall. By way of example, if all plastic bottles used globally were made from glass instead, the additional carbon emissions would be equivalent to powering around 22 large coal-fired power plants. This is equivalent to the electricity consumed by a third of the UK."

But that kind of ludicrous move is what many are advocating for, all because they did not check the facts or because they are so overcome with plastiphobia that they would rather destroy the environment than face facts.

There are multiple LCA studies on drink containers, and they all reach the same conclusion — namely that the PET bottle is the best option and substantially reduces impact compared to glass bottles or metal cans.

Compared with a 20 oz. PET Plastic Bottle	Solid Waste Generated	Energy Expended to Create	Global Warming Potential	Emissions Produced That Contribute to Acid Rain and Smog Formation
12 oz. Aluminum Can	**3x** more	**3x** more	**2x** more	**2-3x** more
12 oz. Glass Bottle	**14x** more	**5x** more	**5x** more	**7-10x** more

"The LCA found that PET plastic bottles, when compared to aluminum cans and glass bottles, are significantly more advantageous for the environment as a beverage delivery system. PET bottles are more sustainable and have a lower impact on several key environmental metrics, including greenhouse gas emissions, expended energy, water consumption, smog, acid rain and eutrophication potential."

Life Cycle Assessment of Predominant U.S. Beverage Container Systems for Carbonated Soft Drinks and Domestic Still Water, Franklin Associates, 2023

"The life cycle environmental impacts of a carbonated drink have been estimated considering four packaging options: 0.75 l glass bottles, 0.33 l aluminium cans, 0.5 and 2 l PET bottles. It has been found that, under the assumptions made in this study, the drink packaged in 2 l PET bottle has the lowest impacts for most impact categories, including global warming potential. Glass bottle is the least preferred option for most impacts."

D. Amienyo et al., Life cycle environmental impacts of carbonated soft drinks, The International Journal of Life Cycle Assessment, 18, pp. 77–92, 2013

Still, on the topic of PET bottles, the study highlighted that improved efficiency in using plastic has dramatically reduced the mass of material needed and therefore its impact.

"Technological advances and changes can also alter LCA results, as materials improve over time. Over the past years the gram weight of the 16.9 ounce 'single serve' bottled water container has dropped by 32.6 %. The average PET bottled water container weighed 18.9 grams in 2000 and by 2008, the average amount of PET resin in each bottle has declined to 12.7 grams."

During my visit to São Paulo in 2024, Plastipak revealed their latest 500 ml water bottle, which weighed only 8 g, a significant reduction from the previous 9 g bottle. There has been a total weight reduction of about 60 % since the year 2000. This is impressive and important, but not something that the public is aware of at present.

The LCA review mentioned earlier noted that only 2 % of the public believed plastic to be the lowest GHG option. People opt for materials that not only increase impact but are often not plastic-free anyway. We all know that paper is not waterproof, so the "paper" cups and cartons are lined with plastic. Metal cans are attacked by liquids, so cans are lined with plastic. We are being sold alternatives that simply make no sense on any level, unless, of course, you are the one profiting from the sale.

"According to a recent YouGov poll, only 2 % of British people consider plastic, compared to other materials used in packaging, to contribute the least greenhouse gases to the environment from its production, use, and post use. The survey findings prompted a better understanding of the issues amongst the wider public to help them make "informed" decisions. Indeed, as reviewed in this work, in terms of carbon emissions, plastic is often the packaging material that is least damaging to the environment from a whole life cycle perspective, particularly when used in closed loop recycling, and most alternative packaging are actually not plastic free."

"If all plastic bottles used globally were made from glass instead, the additional carbon emissions would be equivalent to 22 large coal-fired power plants producing enough electricity for a third of the UK."

N. Voulvoulis et al., Examining Material Evidence – The Carbon Footprint, Centre for Environmental Policy, Imperial College London & Veolia UK, ACC, 2019

Here is another life cycle study on drink containers that was just released recently. It agrees with the prior studies finding that the PET bottle is the best choice.

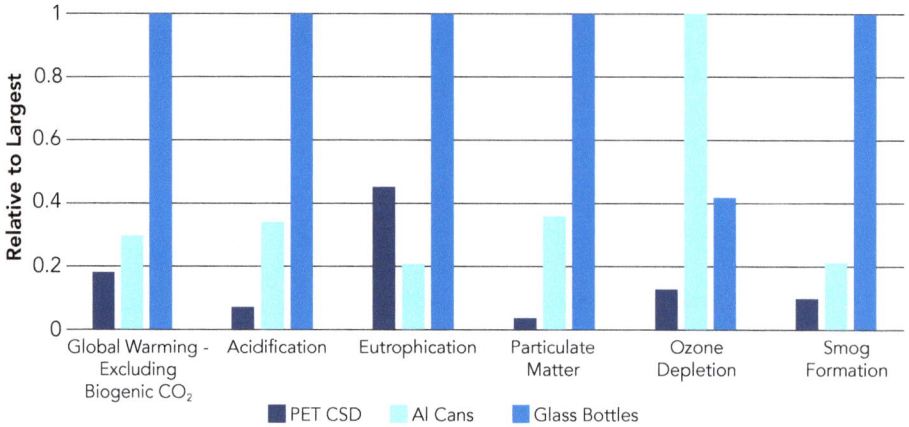

Figure ES-1: Comparison of cradle-to-grave impacts for 500 mL beverage containers in the US (TRACI 2.1).

Comparative LCA on 500 mL Beverage Packaging Products, Sphera™ 2023

When confronted by irrefutable evidence that plastic is usually the option that causes least waste, GHG, fossil fuel use, and total impact, some people then claim that LCA is not valid because it is not perfect or does not include every possible impact factor. This line of thought is not born of a genuine desire to do what is right but, instead, represents a desperate attempt to ignore all evidence so that the person can maintain their anti-plastics stance. Some anti-plastics people are cult-like in their obsession, and no amount of evidence or logic can reach them.

Throwing out LCA, the only proven and effective tool, would be irresponsible. What would we do then? Toss a coin to decide what's greenest? No, LCA works, and it contains all significant factors. In fact, plastic often comes out best in the majority or all factors, so adding a new one would make no difference to the outcome.

Powerful forces are at work to scare us away from the greenest, safe solution, according to virtually every life cycle study and the peer-reviewed science. We are told that PET leaches BPA when there is no BPA in PET and never has been. We are told to be worried about microplastics when, as shown already, they are not actually a problem. Every time you see an attack on PET, it is an attempt to line someone's pockets, not to protect you.

Shopping bags

One very popular topic is shopping bags. I found 24 LCA studies and shared them with an LCA expert to get his professional opinion.

"From all 24 reports and reviews assessed, the actual LCA analyses on grocery bags overwhelmingly point to plastic (HDPE) as the material with least environmental impact, both at single use level and multi-purpose."

Neil Shackelton — Founder Medoola

Here are some quotes from a few of those studies.

Clemson University LCA Study

"Our results also show that Paper bags, even with 100 % recycle content, have significantly higher average impacts on the environment than either of the reusable bags or single-use plastic retail bags"

R. M. Kimmel, Life Cycle Assessment of Grocery Bags in Common Use in the United States, Clemson University, Environmental Studies 6, 2014

UK LCA Study

"The conventional HDPE bag had the lowest environmental impacts of the lightweight bags in eight of the nine impact categories"

C. Edwards & J. Meyhoff Fry, Life cycle assessment of supermarket carrier bags: a review of the bags available in 2006, Environmental Agency, UK, 2011

Franklin Associates LCA

"This study supports the conclusion that the standard polyethylene grocery hag has significantly lower environmental impacts than a 30 % recycled content paper bag and a compostable plastic bag"

Resource & Environmental Profile Analysis of Polyethylene and Unbleached Paper Grocery Bags, Franklin Associates, 1990

Reason Foundation LCA Study

"Unfortunately, policymakers have been cajoled into passing ordinances that ban plastic bags. That is bad news for consumers. It is also bad news for the environment, since the public has been misled into believing that by restricting the use of plastic bags, the problems for which those bags are allegedly responsible will be dramatically reduced."

J. Morris & B. Seasholes, How Green is that Grocery Bag Ban? An Assessment of the Environmental and Economic Effects of Grocery Bag Bans and Taxes, Reason Foundation, USA, 2014

"In general, LDPE carrier bags, which are the bags that are always available for purchase in Danish supermarkets, are the carriers providing the overall lowest environmental impacts when not considering reuse. In particular, between the types of available carrier bags, LDPE carrier bags with rigid handle are the most preferable. Effects of littering for this type of bag were considered negligible for Denmark."

Life Cycle Assessment of grocery carrier bags, Ministry of Environment and Food Denmark, Danish Environmental Protection Agency, 2018

There are now 30 LCA studies on bags right up to the present, and the results are conclusive. Plastic causes the least impact, and it is not even close. Why then are we taxing and banning the greenest choice? Because people are not checking the facts before they act. That is especially negligent because you can type "LCA bag" into Google and find studies in under one second. Anyone who did not manage that amount of effort is not really trying, are they?

There have been a lot of taxes and bans on bags, even though the evidence shows that is a terrible idea. Interestingly, when scientists checked the effect of plastic bag bans, they found not only a vast increase in greenhouse gas (GHG), as predicted, but ironically, an enormous increase in the sales of plastic. How can that be? The reason is that people often reuse their shopping bag as a bin (trash can) liner. However, when the bags are banned, they are forced to buy bin liner bags, but those are made of much thicker plastic. A study found that in the UK, over 75% of shopping bags were reused at least once, usually as a bin liner.

C. Edwards & J. Meyhoff Fry, Life cycle assessment of supermarket carrier bags: a review of the bags available in 2006, Environment Agency, UK, 2011

Freedonia conducted a retrospective study to evaluate the impact of the New Jersey bag ban, and their findings are quite startling.

Bag Demand (million bags)

2015: **2316** (20, 13, 57, 2226)
2022: **894** (130, 80, 395, 289)

Plastic Consumption (million pounds)

2015: **56** (7, 3, 12, 34)
2022: **151** (46, 21, 80, 4)

GHG Emmission from Production (million kg CO₂ eq)

2015: **221** (12, 6, 26, 178)
2022: **311** (75, 35, 178, 23)

Bag Demand CAGR	22/15	Plastic Consumption CAGR	22/15	GHG Emissions CAGR	22/15
Total demand	-61%	Total demand	169%	Total demand	41%
Other plastic bags	550%	Other plastic bags	550%	Other plastic bags	550%
Nonwoven Polypropylene	593%	Nonwoven Polypropylene	593%	Nonwoven Polypropylene	593%
Woven polypropylene	515%	Woven polypropylene	515%	Woven polypropylene	515%
Plastic film bags	-87%	Plastic film bags	-88%	Plastic film bags	-87%

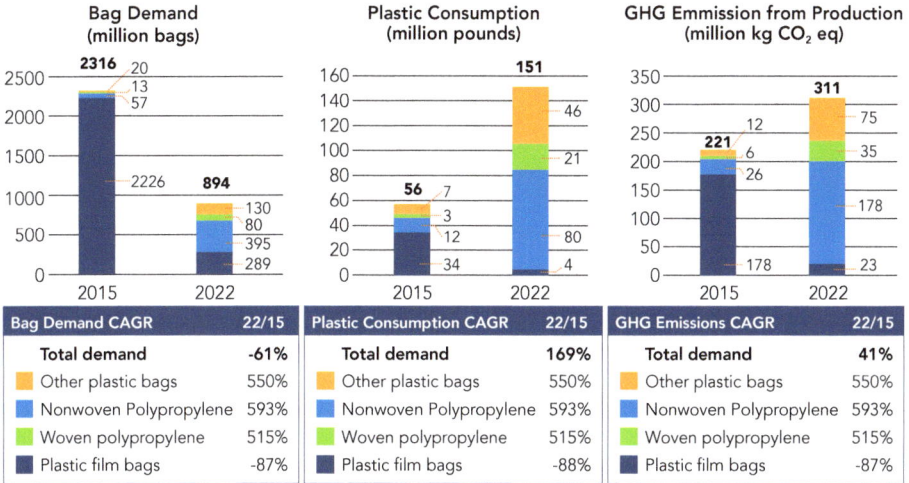

While total bag volumes declined by more than 60% by 2022, the polypropylene plastic consumed to produce NWPP and WPP bags grew by more than 6x. Furthermore, due to the larger carbon footprint of a polypropylene bag, greenhouse gas emissions (ie, CO₂) espanded more than 500%.

New Jersey Retail Bag Market Assessment, Freedonia, December 2023

"As a result, alternative bag sales grew exponentially, and the shift in bag materials has proven profitable for retailers. An in-depth cost analysis evaluating New Jersey grocery retailers reveals a typical store can profit $200,000 per store location from alternative bag sales — for one major retailer this amounts to an estimated $42 million in profit across all its bag sales in NJ."

Greenwashing may be profitable, but it makes matters worse, meaning a large increase in greenhouse gas emissions and, ironically, in plastic sold. According to the data, the plastics industry should be out lobbying for plastic bag bans because they result in increased plastic sales.

Here is another comment about the measured effects of disposable carryout bag (DCB) policies in California. They found that policies against plastic bags increased materials use and greenhouse gas, as one would have predicted from the many life cycle studies.

"This article is the first to evaluate how regulating the use of plastic and paper carryout bags affects the sale of unregulated disposable bags. Using quasi-random variation in local government policy adoption in California in an event study design, I find that the banning of plastic carryout bags leads

to significant increases in the sale of trash bags, and in particular small trash bags. When converted into pounds of plastic, 36% of the plastic reduction from DCB policies is lost due to consumption shifting towards unregulated plastic bags. Moreover, the increase in pounds of paper used from paper carryout bags more than offsets the decrease in pounds of plastic, which has negative implications with respect to the carbon footprint of DCB policies."

R. L. C. Taylor, Bag leakage - The effect of disposable carryout bag regulations on unregulated bag, University of Sydney, School of Economics, 2018

Incredibly, even though all evidence shows PE bags decrease impact and banning them increases it, take a look at this new headline…

"California governor signs law banning all plastic shopping bags at grocery stores"

Associated Press, Published 5:42 PM EDT, Sunday 22nd September, 2024

One wonders if our politicians can read. Such bans eventually get reversed years later when they see the effect, but why do politicians set themselves up to look like fools?

In the UK, you are charged for single-use plastic bags, the ones that cause least impact, but the paper bags that weigh ten times more, create more greenhouse gas, and use more fossil fuel are free. That's insanity. The sales from plastic bags are supposed to go to "good causes." As any wise economist will tell you, keeping the money in your own pocket is the wise and just way to make sure that each of us may decide what a good cause is.

Envelopes

All life cycle studies on envelopes found that PE plastic packaging had a far lower impact than paper-based alternatives.

"In summary, the poly flexible mailer, as well as the bubble mailer made from HDPE, came in with the lowest environmental impacts across a range of metrics, including fossil fuel use, greenhouse gas emissions, water use, material used, and the amount of material discarded."

T. Bukowski, M. Dingee, Sustainability Life Cycle and Economic Impacts of Flexible Packaging in E-commerce, PTIS, LLC, 2021

A major factor is weight. Heavier, paper-based opinions require more gasoline and diesel to transport, which leads to more carbon dioxide compared to the lighter plastic options (because burning fossil fuel creates carbon dioxide).

"The main conclusion that can be drawn from this analysis regarding packaging options for shipping mail-order soft goods to residential customers is that the weight of the packaging is the most critical factor influencing the environmental burdens."

Lifecycle Inventory of Packaging Options for Shipment of Retail Mail-Order Soft Goods, Franklin Associates for Oregon Department of Environmental Quality & US EPA, 2004

With inflatable cushions for packaging, it is the weight of the cardboard box, not those plastic pillows that dominate the impact, so choosing the smallest, lightest viable box is key.

Ironic then that Amazon and Google had press releases to announce that they would move to paper packaging. This is the danger of misleading customer — companies follow their customers' demands even when the choice is detrimental.

Takeaway containers

A life cycle study compared the impact of three material alternatives for takeaway containers. They found that the polystyrene foam clamshell has the lowest impact. The reusable PP container would have to be reused 3–39 times to break even with the exceptionally low impact of the PS foam.

"The best option among the three is the EPS container with the lowest impacts across the 12 categories. Against the aluminium container, its

impacts are 7 % - 28 times lower and against the PP, 25 % to six times better. The EPS is also the best option when compared to reusable takeaway PP containers, unless these are reused 3-39 times, depending on the impact."

A. Gallego-Schmid et al., Environmental impacts of takeaway food containers, Journal of Cleaner Production, 211, pp. 417–427, 2019

Another more recent study also concluded that polystyrene foam has a much lower impact than polypropylene or biodegradable (PLA) food containers. The foam has such a remarkably low impact because it uses so little material, being composed mainly of gas. That means less material used, less waste, less energy, and a lower transportation impact.

"In conclusion, single-use plastic containers manufactured from polypropylene have significant environmental impacts. However, biodegradable containers are not the best alternative, as they have more negative impacts compared to other single-use containers such as styrofoam. Styrofoam is also included in the single-use plastic ban. As such, these results conclude that single-use alternatives do not necessarily have the lowest environmental impacts."

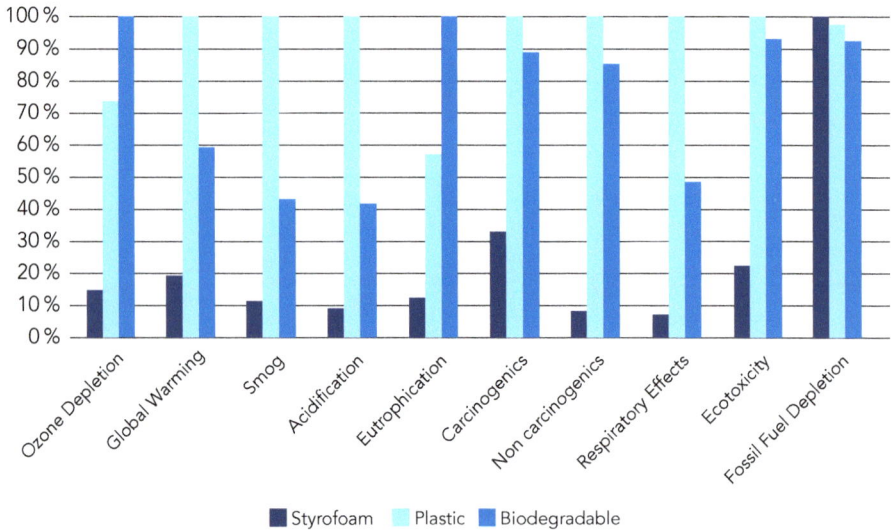

R. Goodrum et al., Life Cycle Assessment of Banned Single-Use Plastic Products and Their Alternatives, Microplastics, 3, pp. 614–633, 2024

The single-use option turned out to have the lowest impact and the material that caused the least impact had been banned without checking the science first. This kind of knee-jerk reaction policy is counterproductive and irresponsible.

Plastic pipes

In 2023, Beyond Plastics released a report claiming that plastic pipes are dangerous and specifically recommending that we use copper pipes instead.

"Lead's impact on our health has been and continues to be horrific. The issue is so significant that in November 2021, Congress made $15 billion available to municipalities to replace lead service lines — a very positive decision that we applaud. But replace these problematic lead lines with what, exactly? While dealing with the lead problem, will we be unintentionally creating new and different problems? After Congress voted to provide this $15 billion, I inquired if they had considered what piping material should be used to replace the lead pipes. The answer was no. I then asked the EPA if it would offer guidance on what material should be used to replace the lead pipes. Again, the answer was no."

"Those two answers inspired the publication of this report."

M. Wilcox, The Perils of PVC Plastic Pipes, Beyond Plastics 2023

Note that they admit to seeing an opportunity to influence where the $15 bn from Congress would go. That, and the fact that they are funded by Michael Bloomberg to attack plastics, should surely raise some suspicions, but their allegations were taken at face value, with no questions from reporters and Congress.

Are they correct in their claims about pipes? What does the science show?

You may recall that earlier we saw plastic pipes are the choice with the least impact, according to a review of life cycle studies. Why then would a so-called environmental group suggest we move to an alternative that increases impact? Let us first look at the LCA data and then at the allegations in their report.

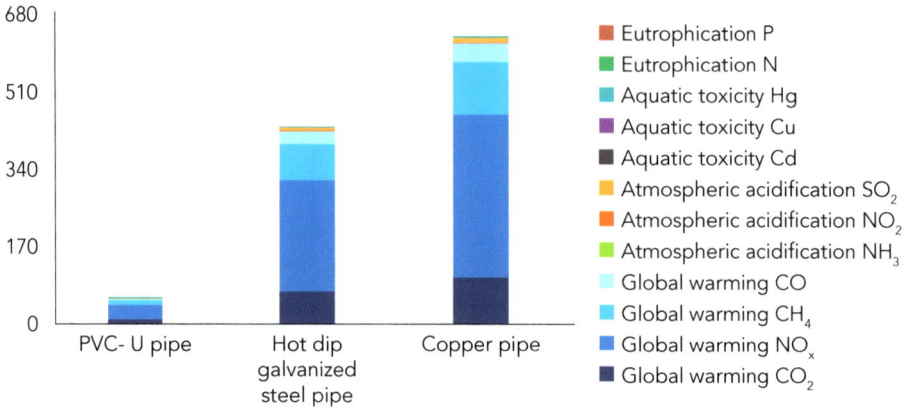

Legend:
- Eutrophication P
- Eutrophication N
- Aquatic toxicity Hg
- Aquatic toxicity Cu
- Aquatic toxicity Cd
- Atmospheric acidification SO_2
- Atmospheric acidification NO_2
- Atmospheric acidification NH_3
- Global warming CO
- Global warming CH_4
- Global warming NO_x
- Global warming CO_2

J. Xiong et al., The application of life cycle assessment for the optimization of pipe materials of building water supply and drainage system, Sustainable Cities and Society, 60, 2020

Reviewing the many life cycle studies on pipes, it is very clear that copper and ductile iron pipes have a much greater environmental impact than plastic pipes made of PVC, PE, PEX, or PP.

HOT AND COLD WATER PLUMBING PIPE LCA
Plastic PEX pipe is far greener than copper

Cross-linked Polyethylene (PEX) pipe systems vs environmental impact comparison
VITO for TEPPFA, 2012

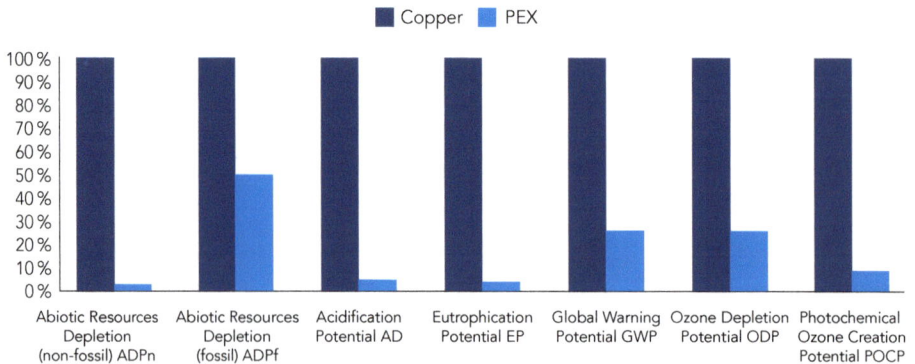

Cross-linked polyethylene (PEX) pipe systems vs copper environmental impact comparison, VITO for TEPPFA, 2012

Do you know what Beyond Plastics said when confronted with the many life cycle studies I provided to the reporter? They told *USA Today* that I was "cherry-picking." That is quite incredible because I shared over 10 life cycle studies while they shared zero then accused me of being selective with data. They were so selective that they showed none at all, perhaps because the facts did not agree with their message? Why let facts and genuine concern for the environment get in the way of a profitable anti-plastics campaign, after all?

Getting back to the report, it made these major claims:

- PVC pipes leach vinyl chloride monomer into the water, presenting a danger.
- PVC pipes leach phthalate plasticisers into drinking water, creating toxicity concerns.
- PVC pipes can create benzene when heated in a forest fire.

They cite a Cornell study, but that study specifically states that vinyl chloride in the water of homes never exceeded the EPA limit even after the water sat stagnant in the pipes for years.

"PVC/CPVC pipe reactors in the laboratory and tap samples collected from consumers homes (n = 15) revealed vinyl chloride accumulation in the tens of ng/L range after a few days and hundreds of ng/L after two years. While these levels did not exceed the EPA's maximum contaminant level (MCL) of 2 mg/L, many readings that simulated stagnation times in homes (overnight) exceeded the MCL-Goal of 0 mg/L."

R. K. Walter et al., Investigation of factors affecting the accumulation of vinyl chloride in polyvinyl chloride piping used in drinking water distribution systems, Water Research, 45 (8), 2011

That same study cited by Beyond Plastics said that vinyl chloride is formed in copper pipes even with no PVC pipe present by a chemical reaction with the chlorine-based disinfectants used. Why did they "forget" to mention that? Is that honest?

"Results from the controlled laboratory experiments with chlorinated and dechlorinated tap water with CPVC and, especially, copper pipes provided the preliminary evidence that VC may be a DBP."

Their selective presentation of information strongly suggests that their goal is not to show risks from vinyl chloride but to attack plastics and unjustly glorify copper.

They also cite a study called "Health Effects from Vinyl Chloride Leaching from Pre-1977 PVC Pipe," which is about pipes produced half a century ago and is not relevant today. The recognition of problems in the past is precisely why they implemented strict regulations that ensure there are no problems now.

R. L. Flournoy, D. Monroe, N.-H. Chestnut & V. Kumar, Health Effects from Vinyl Chloride Leaching from Pre-1977 PVC Pipe, American Water Works Association, 1999

So far, Beyond Plastic's "evidence" has been worthless. Here is another study they cited.

M. Beardsley & C. D. Adams, Modeling and Control of Vinyl Chloride in Drinking Water Distribution Systems, Journal of Environmental Engineering, 129 (9), 2003

That study said that if you have a 50-year-old pipe and a dead end where water stagnates, then you might find higher levels there. Does that present an actual threat to human health, though? No, because those dead ends are where water doesn't flow, so no one can drink it.

Beyond Plastics' "evidence" is just one load of nonsense after the next.

What about their claim that phthalates leach from PVC pipes? PVC pipes are made from rigid uPVC, where the "u" stands for unplasticised, which means no phthalates are in the PVC. How can you possibly get something out of a pipe that was never there in the first place?

They do cite a study mentioning phthalates, but the study has no mention of toxic levels. Just detecting traces of a substance does not mean there is a problem. In fact, detectors are so sensitive now that it is possible to "detect" almost anything almost anywhere.

T. Tomboulian et al., Materials used in drinking water distribution systems - contribution to taste-and-odor, Water Science & Technology, 49 (9), pp. 219–226, 2004

Lastly, they make the incredible claim that if there is a wildfire, they will find benzene in the water; then, they speculate that perhaps it came from the PVC pipe. Only they present no evidence to show that it did, and the science provided contained major flaws so basic that it is hard to believe that professional scientists were involved. Other scientists had no problem working out what had happened regarding benzene detection in wildfires. The fire creates a vacuum in the water lines that pulls in gas from the fire, which is how chemicals get into the water.

"Benzene contamination was present in 29% of service connections to destroyed structures and 2% of service connections to standing homes."

"The fact that concentrations of benzene were highest in service lines to destroyed homes is consistent with the hypothesis that chemical pyrolysis products were pulled into the service lines due to loss of system pressure."

How did those chemicals form? What forms when trees burn in a wildfire? An enormous amount of benzene is created! In fact, each kilogram of wood burnt creates 1 gram of benzene.

"Residential wood combustion is a notable source of benzene, toluene, and the xylenes. Hardwood combusted in the wood stove emits over 1 g of benzene/kg of wood burned."

J. D. MacDonald et al., Fine Particle & Gaseous Emission Rates from Residential Wood Combustion, Environmental Science & Technology, 34 (11), pp. 2080–2091, 2000

Can you believe that it never occurred to the scientists Beyond Plastics cited that the thousands of tons of burning trees in a wildfire might be the source of the benzene they found? Apparently, they were too determined to blame it on plastic. As a scientist, I am profoundly unimpressed.

In 2023, *USA Today* published the nonsense from Beyond Plastics, and I wrote to the journalist explaining that she had been tricked into publishing misinformation. I showed her the evidence; mortified, she instantly offered to publish a correction, which she did a couple of days later. I thought that the matter was closed.

Then, in 2024, when all of that had died down, I started getting a fresh wave of telephone calls from journalists investigating Beyond Plastics' already debunked accusations about plastic pipes. A journalist called from *The Washington Post*, and to her credit,

did her due diligence by checking the claims. She asked whether dangerous levels of vinyl chloride could leach from pipes, and I replied that was indeed a problem pre-1977, but now all plastic pipes are regularly tested, according to NSF/ANSI/CAN 61, which measures for vinyl chloride (down to 0.2 parts per billion concentrations), phthalates, and more. Not only that, but the testing people perform surprise inspections at pipe factories and take samples to test. They have never found a problem.

I sent study after study to prove every point I made, and the journalist concluded it comes down to a "case of he said, she said." I said yes — a group paid to criticise plastics made claims without evidence and a group of respected scientists working unpaid disproved the claims using comprehensive peer-reviewed evidence.

How many times can such groups be allowed to spread nonsense that increases harm to the environment? Journalists really should share a database of discredited sources to save their own reputations and save us from exposure to nonsense.

Perhaps the most worrying part was when I checked the claims that copper is the safer option. Nothing could be farther from the truth. Copper is so toxic that amounts are regulated and tested along with lead by the EPA and other environmental agencies all over the world. The safety limit for its concentration is set at around 1 part per million because of its extreme toxicity.

As the copper pipes are used, they corrode, creating particles and soluble copper salts that are classified as "extremely toxic." The threat is not theoretical either: toxic concentrations are reported in the real world, including in school drinking fountains.

I read over a hundred studies on that, unpaid, and published a report you can find at iscoppersafe.com. Remember, copper is the "safe" choice endorsed by Beyond Plastics. They don't disclose all the groups funding them, but something seems very fishy. Perhaps a journalist should ask whether the copper companies or plumbers' union pays them for their endorsement.

CIRCULARITY

The concept of circularity looks so beautiful — I admit it. Just look at the simplicity of this diagram.

LINEAR ECONOMY   RECYCLING ECONOMY   CIRCULAR ECONOMY

But circularity is not the same as being green, i.e. causing minimum impact. Often, chasing the idealistic dream of circularity means more waste, GHG, fossil fuel burnt, cost, and overall harm. The reasons become clearer when we replace the ideal image that we are always shown with the real diagram that demonstrates all the energy needed to drive the circle and the additional waste streams created.

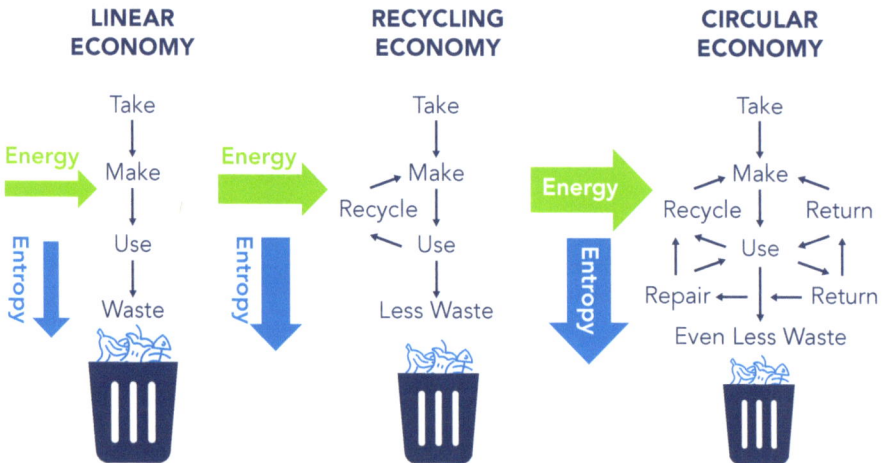

LINEAR ECONOMY

Take
↓
Energy → Make
↓
Entropy ↓ Use
↓
Waste

RECYCLING ECONOMY

Take
↓
Energy → Make
Recycle ↓
Entropy ↓ ← Use
↓
Less Waste

CIRCULAR ECONOMY

Take
↓
Energy → Make ←
Recycle ↓ Return
↑ ↗ Use ↖ ↑
Repair ← ↓ ← Return
Even Less Waste

Redrawn after an original image from Paul Martin of Spitfire Research

It has been pointed out that the circularity concept is largely about morphing a materials problem into an energy problem. Perhaps, one day, we will have unlimited free, green energy; at which point, circularity will become viable, but until then, we need to do the calculations to see whether the circular approach really reduces the overall impact in each case.

"…the circular economy risks turning into a hypothetico-normative (but self-serving) utopia that derails actual and well-intended efforts to reorganize production, consumption, and more generally material flows in ways that are more respectful of planetary boundaries and that work in favor of sustainability."

H. Corvellec et al., Critiques of the circular economy, Journal of Industrial Ecology, 26 (2), pp. 421–432, 2021

"Every loop around the circle creates dissipation and entropy, attributed to losses in quantity (physical material losses, by-products) and quality (mixing, downgrading). New materials and energy must be injected into any circular material loop, to overcome these dissipative losses."

"For the most part, the novel solutions the CE purports to provide in the handling of materials merely shift impacts to the energy domain."

J. M. Cullen, Circular economy: Theoretical benchmark or perpetual motion machine? Journal of Industrial Ecology, 21 (3), pp. 483–486, 2017

Cullen calculated the circularity index of concrete to be zero, meaning that there is no energy saving from recycling it compared to making new concrete. As concrete makes up most of the materials we use, that means a large portion of material is not worth recycling, i.e. the circularity concept fails to deliver a benefit.

Steel and aluminium have a high positive circularity index, so they are well worth recycling. Paper and plastic both have a low positive index, so there is a benefit to recycling them, although far less so. Later, in the book, there is a table showing that expensive materials tend to be worse for the environment and cheaper ones are less harmful. This is why it does not make sense to move from paper and plastic to metal or glass — it increases cost and impact even though those materials are more attractive to recycle from a cost and energy perspective.

When I worked for Electrolux/Frigidaire, a sustainability manager told the Board that we should move from plastic washing machine tubs (the part inside that holds the water) to steel ones, because at the end of life, the steel one has more value. We now understand why he was wrong and had to backtrack on his policy suggestion.

Do you want to chase a dream or make the real world a better place?

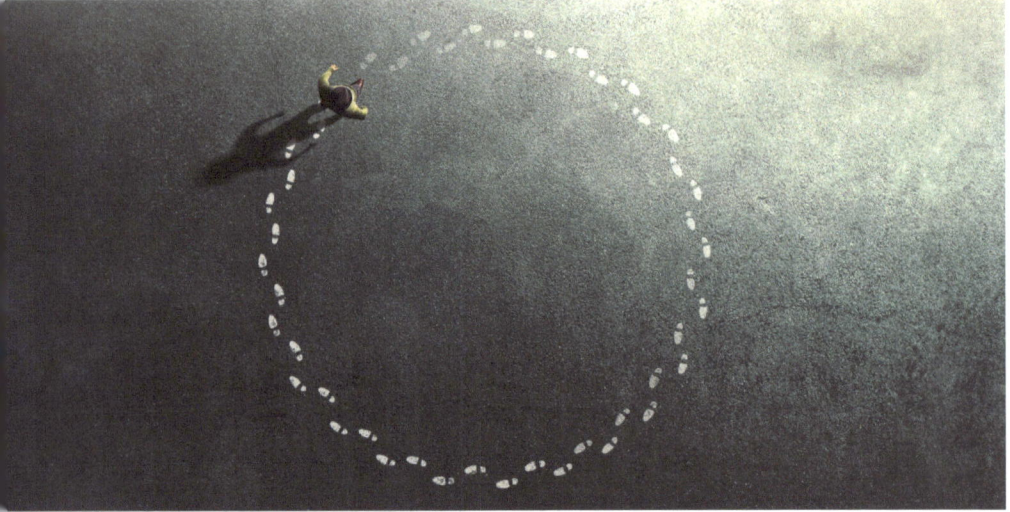

Look at the man walking around in circles — he completes the circle, but it costs energy, he needs to eat, and it creates waste (toilet breaks). Circles have costs.

Circular economy practices, defined as the ten Rs (Refuse, Rethink, Reduce, Reuse, Repair, Refurbish, Remanufacture, Repurpose, Recycle, Recover) often make sense, but in each case, we need to double-check the facts to be sure.

SINGLE-USE

There is so much focus on single-use and how it is so bad for the environment. Therefore, we should look at that in more detail and see what the facts are.

This study found that reuse is not the panacea we have been led to believe it is. There are trade-offs because reusables need to be more durable, requiring more material and more impact to create. So, they increase impact unless they really are reused enough times.

"Substituting single-use plastic for other single-use materials does not represent a solution in most cases. LCAs of single-use packaging tend to show plastics having the lowest impacts, mainly as a result of plastics' low weight relative to other materials. Substitutions with cardboard, glass, steel or aluminium tend to show higher impacts, or at best, a trade-off between different impacts."

Single-use supermarket food packaging and its alternatives: Recommendations from Life Cycle Assessments, UNEP, 2022

What can LCA tell us about single-use cutlery on airlines? It would be natural to assume that reusables must be a better alternative, and I see people post about just that.

"The paper reveals that the lighter single-use packaging and tableware for airline catering are less harmful under a life cycle perspective…"

"In cases where transport is the dominant stage, as in aviation, it can be observed that much lighter single-use items generate less greenhouse gases throughout their complete life cycle."

G. Blanca-Alcubilla et al., Is the reusable tableware the best option? Analysis of the aviation catering sector with a Life Cycle Approach, Science of The Total Environment, 708 (15), 2020

Are you surprised by the result? The reason is that any increase in weight on an aircraft means more fuel burnt and so more carbon dioxide created by the combustion of the fuel. This example highlights why going with our "gut" or intuition can lead to poor choices.

A Life Cycle study found that single-use PET bottles have lower impact than reusable PET bottles and both types of plastic bottle had lower impact than reusable glass beverage bottles.

H. Lerche Raadal et al., Life cycle assessment of the current recycling system and an alternative reuse system for bottles in Norway, Norwegian Institute for Sustainability Research (NORSUS), Report OR.27.23, 2023

When it comes to reuse, I have observed a common misconception. Namely, people are keen to ditch single-use plastic items like containers and move over to multi-use metal or glass options. As we have established, reusables are often greener, so there is nothing wrong with the idea of making a shift, but why a shift to metal or glass when both are vastly worse for the environment and more expensive as well? It never occurs to most people that the lower-impact and less expensive option is a reusable plastic container. Either there is a logical breakdown in the minds of these customers, or their true goal is not to avoid single-use products but rather to go buy metal and glass.

I can completely understand the desire to go for metal and glass from an aesthetic perspective, for the touch, for the perception of quality… I too am attracted to such products. However, people should be aware that they are deceiving themselves if they think it will help the environment, and many companies are happy to help deceive you with false green claims to get your money out of your pocket.

After seeing the objections to single-use, upon reflection, I am not so certain that single-use is truly what people are against. Let me explain.

Apples are single use and so are many other items. I can only eat an apple once, and then it is gone, but no one minds that single-use application. This gives us one clue.

What about the huge amounts of zero-use paper called spam? After all, zero use is much worse than single use.

"According to its own 2018 annual report and website, the USPS hauled 77+ billion pieces of junk mail across 1.4 billion miles."

"…this still means that 95% of direct mail misses its mark — and is duly tossed out. At a tremendous expense to all of us."

PaperKarma and United States Postal Service

We receive spam in our letterbox and march it right to the bin without even opening it. If single-use is so wasteful and objectionable, then people should be marching in the streets protesting against the environmental atrocity of zero-use paper spam. But they are not, are they? This is another clue that single-use products may not be the actual issue.

When we look at the clues, we conclude that litter is the real issue, not single-use. We do not mind the apple because it is gone once eaten, so no litter is created.

We do not mind the zero-use spam because it goes directly into the bin and is therefore not littered.

It turns out that we are against single-use items only because they are objects so cheap that we can afford to be careless with them and litter them.

Fortunately, we already know the solution for litter — education, deposits, and fines.

Remember, though, that these measures must be applied equally to all materials because imposing deposits and fines only on plastics drives people to paper and other alternatives that lead to more waste, more litter, more impact, and higher cost.

One last thought on straws and single-use. The least impact comes from taking no straw at all. Just say, "No, thanks." The second-least-impact option is the plastic straw, then reuse it as many times as you can. Remember, no one forces us to throw away that straw. It is only a single-use product if we want it to be. People have reused plastic straws 50 or 100 times, and they can be cleaned in the dishwasher. Best of all, the impact is less and less with every re-use. I have "single-use" plastic cutlery that has been used over a hundred times.

MISLED CONSUMERS

My keynote talk is called "The Great Plastics Distraction" because people are so busy obsessing about plastics,

which create around 1% of impact, that they are ignoring the 99% of materials that create vastly more impact. We have zero chance of solving a problem by ignoring 99% of it. Other scientists agree.

"The global discourse surrounding plastics has been marked by a profound perceptual schism, also for plastic packaging in the fresh food industry. The public opinion expresses mounting concerns in terms of such plastic packaging solutions. However, in many cases the unique material properties and the well-established methodology of Life Cycle Assessment (LCA) actually demonstrate the environmental advantage of plastics for food packaging. This paper delves into the chasm between the two perspectives, leveraging empirical evidence to resolve the divide."

E. Horsthuis et al., Closing the Perception-Reality Gap for Sustainable Fresh Food Plastic Packaging, Procedia CIRP, 122, pp. 647–652, 2024

Another team of scientists had this to say about the misguided focus on plastics to the exclusion of everything else.

"Antiplastic sentiments have been exploited by politicians and industry, where reducing consumers' plastic footprints are often confused by the seldom-challenged veil of environmental consumerism, or 'greenwashing.' Plastic is integral to much of modern day life, and regularly represents the greener facilitator of society's consumption."

T. Stanton et al., It's the product not the polymer: Rethinking plastic pollution, WIREs Water, 8 (1), 2021

They went on to say:

"Influenced by media and political exploitation of an emotive environmental issue, public concern for the environment is dominated by plastic pollution However, as a scientific community, it is important that the amount of time

and funds devoted to addressing this popular concern are not disproportionate to less tangible anthropogenic pressures on our environment such as that of heavy metals, pharmaceuticals, and pesticides. Environmental research that does not fairly represent the problem under investigation risks undermining public and political trust in environmental science."

It is correct for them to say that professional, ethical scientists have a duty to report fairly the threats and solutions rather than take the easy route of demonising plastics to the detriment of the community. Currently, our funds and policies misallocate resources if effective environmental preservation is the goal.

SUMMARY

Life cycle analysis is the only proven method that reliably provides an answer to the question of what causes more impact and what causes less impact. While carbon dioxide (GHG) may be the most important factor for many people, plastics usually reduce not only GHG but also material use, waste creation, fossil fuel consumption, toxic effects, and more. Replacing plastic with alternatives increased GHG in 93% of applications studied. Therefore, choosing the plastic option is usually the wisest choice if the goal is to minimise environmental impact.

The good news is that usually, the alternative with the least impact is also the least expensive because impact and cost both depend on the energy used, transportation, weight, water consumption, and so on. So, rather than worrying about how you can afford to go green, which is what people wonder about now, you can pick the lowest-impact option and save money at the same time. That is what you get for having the wisdom to check the facts before you act. Rather than being guilted into spending more on some new "alternative" product like goose eggs or some other in vogue nonsense, you can make a sound choice based on facts and evidence.

Materials generate a significant fraction of greenhouse gas (~25%), but most of that comes from iron, steel, and concrete use — not plastics.

Zooming out from materials alone to the bigger picture, the best way to reduce total impact is to buy less, use less, and act responsibly through the ten Rs:

Refuse, Rethink, Reduce, Reuse, Repair, Refurbish, Remanufacture, Repurpose, Recycle, Recover

PLASTICS RECYCLING - MYTHS & FACTS

PLASTICS RECYCLING: MYTHS & FACTS

As we have just seen in the last chapter, based on life cycle studies, plastic is usually the lowest impact option even with low or no recycling. Having said that, recycling does offer the opportunity to further reduce impact, so it is a topic well worth exploring.

As exposed in *The Plastics Paradox* book, most of what we have been told about plastics and the environment is simply untrue, meaning that decades of science say the opposite. So, let us examine these common assertions about plastics recycling and see what science says about them.

PLASTICS RECYCLING IS NEEDED TO PREVENT LITTER & POLLUTION

We hear that increasing the recycling rate will solve the problem of "plastic pollution." Again, scientists have revealed the facts. It turns out that what many are calling plastic "pollution" is really litter. Whereas pollution is associated with companies, litter is caused by people, and the solutions to that involve changing the behaviour of those people via education, deposits, and fines.

E. Carpenter & S. Wolverton, Plastic litter in streams: The behavioral archaeology of a pervasive environmental problem, Applied Geography, 84, pp. 93–101, 2017

Will increased recycling really help to reduce litter? While there is no evidence that people litter less when a product is recyclable, often recycling does lead to less litter indirectly. It is common to impose a deposit on items; that leads to a large decrease in litter because, once the product has value, due to the deposit, people do not drop it anymore, or if they do, someone else will pick it up to collect the deposit.

A good analogy here is plastic banknotes. They print billions of plastic banknotes every year — how many do you see in the streets, floating down rivers, or on beaches? We never see them littered because although they are small and easily lost, they have value, so people take care of them.

RECYCLING IS NEEDED TO MAKE PLASTICS GREEN

One of the most common claims is that we need to recycle plastics at a much higher rate for plastics to become truly green. Scores of life cycle studies spanning decades show plastics cause the least impact. Replacing them with alternatives like paper, cotton, metals, or glass increases harm, not only in terms of greenhouse gas but also waste created, fossil fuel used, and total impact across all the factors included in modern life cycle studies.

Some of those life cycle studies also ran scenarios assuming different recycling rates for plastics and other materials. As previously mentioned, they concluded plastics create less impact even with low or no recycling at all. So, it is not correct to say that we are waiting for recycling to make plastics the right choice for the environment.

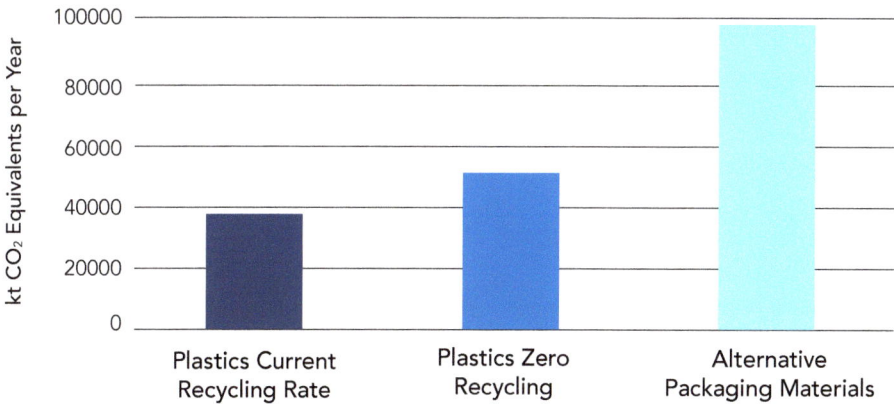

Denkstatt The impact of plastic packaging on life cycle energy consumption and greenhouse gas emissions in Europe, Executive Summary July, 2011

Even so, it is correct to encourage recycling because recycled plastic needs far less energy and creates far less greenhouse gas than new plastic does. Typical reductions are 70–80%, and that is achieved using the standard, cheap, and simple method known as "mechanical recycling." That entails collecting the plastic, separating, washing, shredding, and remoulding it into a new product.

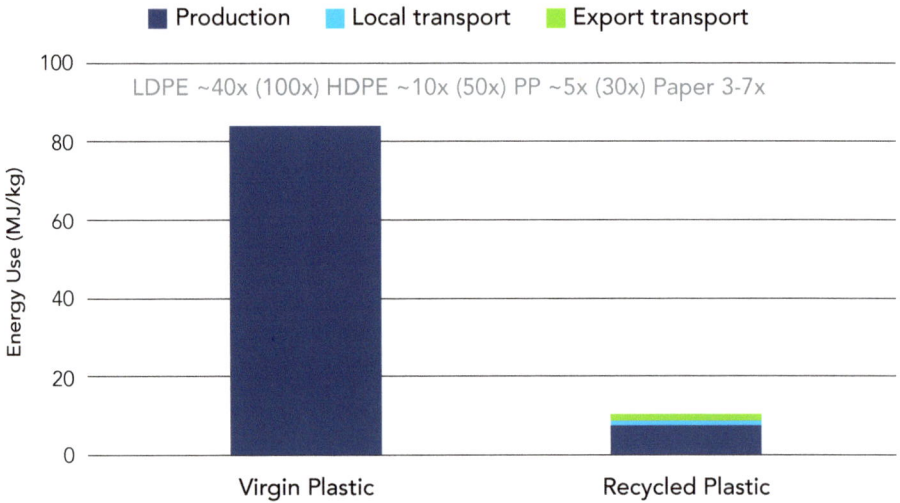

Life Cycle Impacts of Plastic Packaging Compared to Substitutes in the United States and Canada, Franklin Associates for The Plastics Division of the American Chemistry Council, 2018

C. Wong, A Study of Plastic Recycling Supply Chain, University of Hull, 2010

THE PLASTICS CAN ONLY BE RECYCLED ONCE MYTH

You may have heard that plastics can only be recycled a couple of times, whereas metal and glass can be infinitely recycled, so we should pick metal and glass over plastic. Is that correct? No, it is not. In fact, plastic can be recycled many times with good properties, as shown by multiple peer-reviewed studies.

Here is a quote from a study in which they put LDPE through an extruder to recycle it one hundred times. They found good properties until the fortieth cycle and said:

"With these results on hand, it could be concluded that LDPE could be extruded for up to 40 times without significantly changing its processability and long-time mechanical properties"

H. Jin et al., The effect of extensive mechanical recycling on the properties of low density polyethylene, Polymer Degradation and Stability, 97, pp. 2262–2272, 2012

Plastic pellets (sometimes incorrectly referred to as "nurdles" by NGOs)

Similar studies have shown that other common plastics like HDPE, PP, and PET can be recycled multiple times with good retention of properties.

A. Boldizar et al., Simulated recycling of post-consumer high density polyethylene material, Polymer Degradation and Stability, 68 (3), pp. 317–319, 2000

R. Mnif & R. Elleuch, Effects of reprocessing cycles and ageing on the rheological and mechanical properties of virgin-recycled HDPE blends, Matériaux & Techniques 103, 704, 2015

M. Mihelčič et al., Influence of Stabilization Additive on Rheological, Thermal and Mechanical Properties of Recycled Polypropylene, Polymers, 14 (24), 5438, 2022

B. von Vacano et al., Elucidating pathways of polypropylene chain cleavage and stabilization for multiple loop mechanical recycling, Journal of Polymer Science, pp. 1–10, 2023

As for the claim that metal and glass can be infinitely recycled, we know that no material can be recycled indefinitely because of contamination and losses during the process. Such losses are well documented for both metals and glass.

SOFT PLASTICS LIKE LDPE CANNOT BE RECYCLED MYTH

How can mechanical recycling be so widely applicable when we know soft plastics like bags cannot be recycled that way and that black or coloured plastics cannot be recycled mechanically? The simple answer is that those claims about recyclability are also untrue.

Soft plastic, like the low density polyethylene from shopping bags, can be recycled and are recycled. One company in Germany, Papier-Mettler, has recycled over 100,000 tons a year profitably for years, and they are not alone. Other companies have done so as well.

Soft plastic can cause jams in machines not designed to handle them, so rather than install the correct machines, many just label them as "unrecyclable," even though this is not true. More responsible companies invest in equipment to solve the problem instead.

BLACK AND COLOURED PLASTICS CANNOT BE RECYCLED MYTH

Black and coloured plastics have also been labelled unrecyclable when, in fact, they can be recycled perfectly well. It's just that some prefer not to have to deal with them because the resale value of coloured plastic is lower. So, rather than recycle them, some companies mislead the public by calling such materials "unrecyclable."

Many years ago, it was found that the most common black colourant, carbon black, prevented plastics from being sorted for recycling automatically because that pigment confused the detectors. However, that problem was solved long ago by finding black colourants that do not interfere with sorting. I still see claims that black plastic cannot be sorted and recycled, but that is not the case.

What about coloured plastics? Sprite recently removed the iconic green colour from their PET bottles, so now they are colourless. Why? Because there is more demand for colourless recycled plastic, so the market value is higher. That's why colourless plastics are preferred for recycling. Both coloured and uncoloured are equally recyclable, but let's talk about what recyclable really means.

THE DOWNCYCLING MYTH

It is said that plastics cannot be recycled back into the same product again and again but must be made into other, lower-value products. Is that really the case, though? Can plastic be recycled into the same product, and do we always need to see the creation of different products as a negative?

PET bottles are a good example highlighting that yes indeed, bottles can be returned and recycled into new bottles not just once or twice but ten times or more. Then, when the plastic is no longer suitable for bottle-making, it can be spun into fibres and made into a fleece sweater instead. Some claim recycling plastics to make a different product is "downcycling," but that line of thought baffles me. How can anyone believe that turning a cheap soda into a luxurious and durable fleece is downcycling? That seems like a breakdown in reasoning to me. This recycling of PET is not theoretical either; it has been done in enormous volumes in multiple countries for many years. By implementing a deposit system, return rates are extremely high, above 95%.

PET may be one example, but is it an outlier? What about other common plastics?

LDPE has been recycled 100 times with good retention of mechanical properties for the first 40 times. HDPE was recycled 50 times with good properties for the first 10 cycles. Polypropylene was reprocessed 50 times but was not of sufficient quality after the first 10. The idea that plastics cannot be recycled, or can only be recycled once at best, is simply not true.

H. Jin et al., The effect of extensive mechanical recycling on the properties of low density polyethylene, Polymer Degradation and Stability, 97, pp. 2262–2272, 2012

N. Benoit et al., High Density Polyethylene Degradation Followed by Closed-loop Recycling, Progress in Rubber, Plastics and Recycling Technology, 33 (1), 2017

M. Mihelčič et al., Influence of Stabilization Additive on Rheological, Thermal and Mechanical Properties of Recycled Polypropylene, Polymers, 14 (24), p. 5438, 2022

It should be mentioned that the number of recycling cycles can be increased by adding more stabiliser, better stabilisers, and other additives to protect the polymer chains and to repair the damage done. Tiny amounts of additive can give meaningful improvements, and advances in the field continue.

PVC can be and is recycled in large volumes. According to the Vinyl Institute, over 1 billion pounds of PVC is recycled every year in the US and Canada. It is surprising to many that one of the lower-impact materials is inherently flame retardant, highly durable, and non-toxic.

"The work presents a literature review on mechanical and feedstock recycling. The advantages and disadvantages of various recycling methods and their development perspectives are presented. The general characteristics of PVC are also described. In conclusion, it is stated that there are currently high recycling possibilities for PVC material and that intensive work is underway on the development of feedstock recycling. Based on the literature review, it was found that PVC certainly meets the requirements for materials involved in the circular economy."

K. Lewandowski & K. Skórczewska, A Brief Review of Poly(Vinyl Chloride) (PVC) Recycling, Polymers, 14, pp. 3035, 2022

For comparison, paper can only be recycled between 3 and 7 times because the fibres are broken down with every cycle until they can no longer make paper of sufficient strength. We see it claimed that aluminium and glass are green because they can be recycled "infinitely," but as mentioned, that is not true either.

In Norway, they recycle 60–70% of expanded polystyrene foam and are aiming for 90%. EPS is also recycled in large volumes in many other countries.

"In case the milk bottles are removed from the PE sorted product and they are mechanically recycled, a high quality of recycled HDPE can be obtained, which contains less contaminants than the freshly produced milk bottle. The composition of these contaminants is, however, different. In the freshly produced milk bottle only degradation products of HDPE and the antioxidant can be found, whereas in the mechanically recycled HDPE also traces of volatile contaminants are found that originate from the milk, the other packaging components, other packages and the surrounding atmosphere."

E. U. Thoden et al., Volatile organic contaminants in HDPE milk bottles along the mechanical recycling value chain, revealing origins and contamination pathways, Journal of Cleaner Production, 459, 142571, 2024

The quality and purity are so high that several recycled plastics have approval for use in contact with food for which stringent testing is mandated. That includes the most common plastics, like PE, PP, and PET, among others.

WHAT THE WORD "RECYCLABLE" MEANS AND DOES NOT MEAN

The word "recyclable" is in the dictionary; it literally means "able to be recycled."

This familiar symbol is used to indicate whether the material can be recycled.

Plastics are recyclable, and they remain recyclable, whether or not they actually get recycled. So-called environmental groups have launched legal battles over this. They asserted the customer was misled by claims that the product was "recyclable" when in reality, the probability that it would be recycled was low, even though both the dictionary and science say otherwise.

Whether something can be recycled is called recyclability.

Whether something is likely to be recycled in that particular locality is another concept and requires its own word, for example, "recycle-likely."

Since so many people struggle with this idea, here is an analogy.

A football is "kickable," and it remains kickable whether or not we choose to actually kick it.

I order food at a restaurant.
The food remains edible, whether or not I decide to eat it.

The same concept applies to recycling, where NGOs claim that the word recyclability actually means recycle-likely. Then, they complain that companies saying that products are "recyclable" are misleading us because the word is not being used according to their own made-up definition. If they want a word that means recycle-likely, then they should propose a new word and put that in the dictionary, not hijack another word and use that.

There is another flaw in the idea of labelling products with a recycle-likely symbol. What local authorities choose to recycle is up to them and varies widely. It is not possible to say whether that product is likely to be recycled where you choose to dispose of it. What if I buy it in Michigan, then cross over to Ohio and their government has decided not to recycle that product? The same applies to country borders. Many products are made in one country and sold in another, but somehow, the NGOs demand that the manufacturer become clairvoyant and anticipate the recycling policies of the region in which the product is eventually disposed. Sounds unfair to me.

Unfortunately, even the organisations responsible for recycling standards are adding to the confusion. Their goal is to make recycling easier for their members, so they also label non-ideal materials as non-recyclable. This is counterproductive and misleading. See the Association of Plastic Recyclers (APR) for specifics.

IS PLASTICS RECYCLING A SCAM?

Lastly, some so-called environmental groups have made the accusation that recycling is a scam and that it can never work. We now know that to be false. Such groups are known to make up such stories to make people angry enough to donate — Dr. Patrick Moore, the former President of Greenpeace, said so himself.

Greenpeace wants a piece of your green - An independent report by Dr. M. Connolly, Dr. R. Connolly, Dr. W. Soon, Dr. P. Moore and Dr. I. Connolly, December 2018

There is room to improve, especially in the US, where rates are much lower than those in Europe, for example. For the US to catch up, the country needs better collection and infrastructure for sorting and recycling.

The Circular Economy for Plastics: A European Analysis, Plastics Europe, March 2024

Another reason that plastics recycling rates are lower than for some materials is to do with profitability. Expensive materials like platinum, palladium, and gold are terrible for the environment. For example, 27,000 kg of carbon dioxide are created for every 1 kg of gold made. Plastics are the opposite, i.e. they have a very low carbon footprint and are very cheap.

Material	Footprint kg/kg	Price $/ton	Recycling %
Gold	27,000	85,000,000	86
Platinum	15,000	30,000,000	60
Silver	100	1,000,000	50
Nickel	12	15,000	60
Aluminium	12	2500	42
Copper	4	9000	46
Plastic	2-3	1000-2000	10
Paper	0.7	1000-2000	60*
Wood	0.4-0.6	700	15
Concrete	0.12	60	40
Limestone	0.02	35	NA

* around 50% of paper is downcycled into cardboard

Materials and the Environment: Eco-Informed Material Choice 3rd Edition, Michael F. Ashby, Butterworth-Heinemann / Elsevier, Oxford, p. 232, UK, 2021

International Energy Agency, End-of-life recycling rates for selected metals, April 2021

Being cheap means that people litter materials like plastic and paper, so collection rates suffer. Plus, it is not that easy to recycle inexpensive materials profitably because margins are lower. So, far from being a conspiracy, the lower recycling rate is at least partly because of economics. The price of recycled plastic fluctuates wildly, and companies frequently go out of business due to those swings. While plastic can and is recycled profitably, it is not trivial to make it work profitably in the long term. Part of the solution for that is for large companies to sign long-term contracts to buy post-consumer recycled plastic (PCR) at a fixed price. That way, the recycler can be assured of steady business. Another reason for the low recycling rate of plastic is the wide variety of plastic materials and the need to clean and separate the different types before recycling.

NGOs want us to use materials like aluminium and glass, which are much worse for the environment because they are more expensive and therefore more likely to be collected, with the profits from recycling them also being greater. I wrote an article to explain just how bad and illogical that idea is. If your friend told you to buy a Ferrari for $200,000 instead of a $20,000 Fiat because the trade-in value of the Ferrari will be higher when you decide to sell, would you fall for that terrible advice? I hope not.

ADVANCED RECYCLING OR CHEMICAL RECYCLING

You may have seen that there are huge, highly funded projects to create new types of recycling. These so-called advanced recycling methods, such as chemical recycling (breaking the polymer down into its starting materials), or dissolving the plastic in solvent, or pyrolysis, where the plastic is heated and converted into oils or monomers (the building blocks of plastics).

The perception is that we are waiting for advanced recycling to make plastics green, when in reality, standard mechanical recycling works just fine for about 90% of the plastic types we use, such as polyethylene, polypropylene, PET, and PVC. These other more expensive, more complex approaches to recycling may eventually have a place in the future, but they are not the key to success.

L. Veillard, Fifty years: chemical recycling's fading promise: Industry Landscape Overview, Zero Waste Europe, November 2024

Mechanical recycling is proven to be cheap and the best environmentally speaking. Plus, it uses standard machinery already installed all over the world because those machines, called "extruders," are used to process new plastics too.

These more difficult forms of recycling may have a place for the minority of plastic that cannot be mechanically recycled and for plastics that have been mechanically recycled repeatedly until their mechanical properties have declined too much. Even then, however, it may make more sense simply to burn the plastic to create electricity, replacing the need to burn oil, coal, or gas, thus saving fossil fuel.

T. Uekert et al., Technical, Economic, and Environmental Comparison of Closed-Loop Recycling Technologies for Common Plastics, Sustainable Chemistry & Engineering, 11, pp. 965–978, 2023

PYROLYSIS IS A GREEN WAY TO RECYCLE PLASTICS

What about pyrolysis as a way to deal with used plastic? Life cycle studies on pyrolysis reveal that it does not make environmental sense.

Pyrolysis means heating substances without oxygen to convert them into organic liquid or fuel, but plastics are already as energy-rich as oil or coal.

Solid plastic waste can be burnt to create electricity, thus reducing our need to burn fossil fuels like oil, gas, and coal. So, why use pyrolysis to turn solid plastic fuel into a much smaller quantity of liquid fuel? Think of the analogy of changing money. If I have a dollar bill and ask for change but only get 50 cents in coins for my dollar, that would be a terrible deal. That's the same deal on offer with pyrolysis.

"The catalytic pyrolysis of PS produced the highest liquid oil (70 and 60%) compared to PP (40 and 54%) and PE (40 and 42%)..."

R. Miandad et al., Catalytic Pyrolysis of Plastic Waste: Moving Toward Pyrolysis Based Biorefineries, Frontiers in Energy Research, 7, 2019

Pyrolysis is not green and is only researched because people take government money, meaning our tax money, to do it.

The same for other approaches, like dissolving the plastic in solvents or using enzymes to break the plastic down into new monomers. While technically feasible, these methods usually turn out to be red herrings when one considers the investment needed and the impact of the process itself. Why then are there so many headlines and projects on them? That's because people will do whatever they can get funded to do, whether or not it actually makes sense. Some have criticised attempts at advanced recycling, and they have a point. Spending time and money on technologies that do not make sense only increases environmental impact.

SUMMARY

Like all materials and everything we do, plastics have an impact. However, decades of life cycle studies agree that plastic is almost always the alternative that minimises material use, waste, greenhouse gas, fossil fuel use, and total impact. Recycling works and rates are high in many countries; the USA is anomalously low and working to improve. Mechanical recycling is cheap, proven, and works with existing equipment. Let's be wise, which means picking the option that minimises impact, then reuse and recycle it.

We have learnt that the more impact a material has, the more expensive it is, and therefore, the more economically attractive to recover and recycle it is at the end of use. Such high-impact materials may be attractive for recycling, but that is not a reason to choose them. We are told to choose aluminium cans because their recycling rate is high, but that argument is false, though promulgated by companies trying to sell you a product. On the contrary, the wise choice is the material with the lowest environmental impact, and in most cases, that material will also save the customer money, as well as saving the environment.

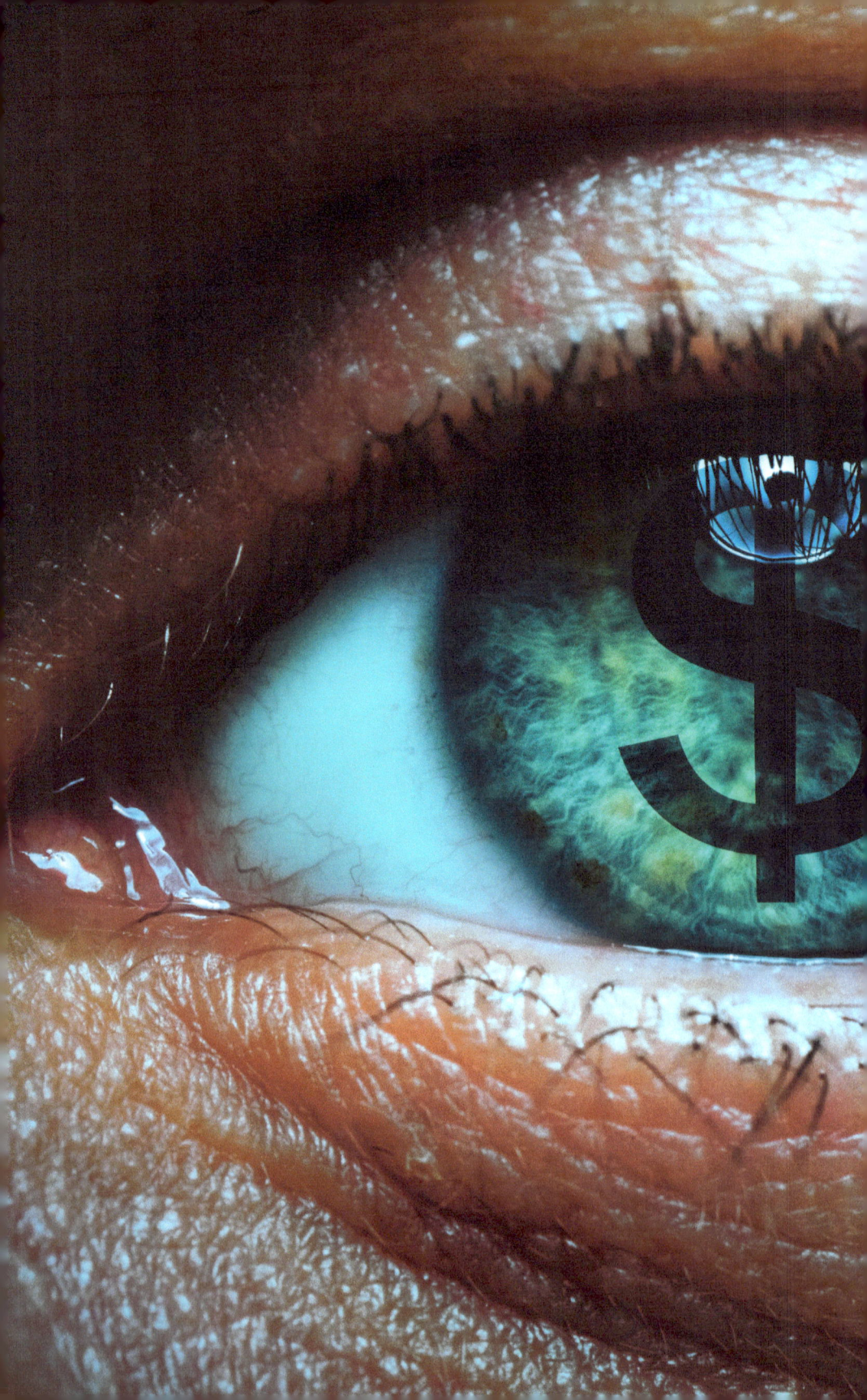

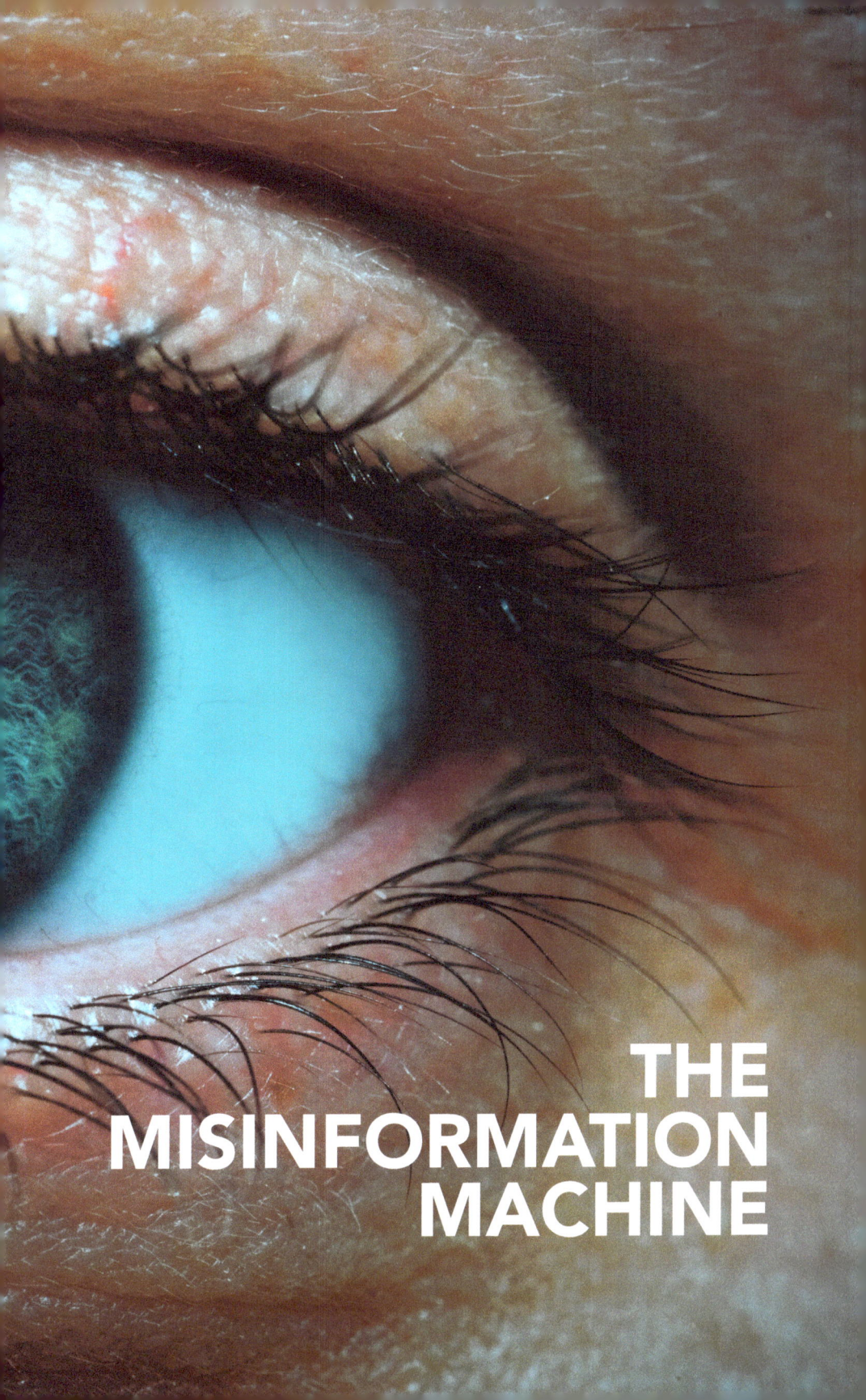

THE
MISINFORMATION
MACHINE

THE MISINFORMATION MACHINE

In this book, we have compared what we are told by so-called "environmental" groups and the media to what peer-reviewed science has to say, and there has been a clear trend, as we can see in this summary.

Topic	NGOs say problem is	Science says problem is	NGOs right or wrong?
Materials	Plastic	Concrete, Wood, Metals	Wrong
Waste	Plastic	Manufacturing, Mining, oil, gas	Wrong
CO_2	Plastic	Metals, cement, paper	Wrong
Fossil fuel	Plastic	Iron, steel, cement	Wrong
Ocean plastic	Dangerous Increasing	"Negligible" Constant	Wrong
Turtles	Plastic	Trawling, fishing, boat strikes	Wrong
Whales	Plastic	Fishing gear, vessel strikes	Wrong
Birds	Plastic	Buildings, powerlines, cats	Wrong
Particles	Plastic	Soot, inorganics (quartz, Pb, Cd)	Wrong
Toxins	Plastic	Lead, mercury, cadmium, dioxins	Wrong

The claims made by the NGOs have been wrong every time. If we were to ask a monkey to guess instead, then statistically, the monkey would do vastly better than these NGOs, who shout their warped messages at our teachers, our children, and our policymakers.

We might wonder whether these NGOs are evil, making up nonsense to relieve us of our money, or simply incompetent. It turns out that there is no need to wonder. The former president of Greenpeace became so disgusted at what they had morphed into that he left and exposed them. According to him, their business model is to make up crises that do not exist to get our money out of our pockets and into theirs. He goes into detail about how they cunningly and systematically implement their strategy, including the attack on plastic materials.

Quote from their former President

Greenpeace is a very successful business. Their business model can be summarized as follows:

- Invent an "environmental problem" which sounds somewhat plausible. Provide anecdotal evidence to support your claims, with emotionally powerful imagery.

- Invent a "simple solution" for the problem which sounds somewhat plausible and emotionally appealing, but is physically unlikely to ever be implemented.

- Pick an "enemy" and blame them for obstructing the implementation of the "solution". Imply that anybody who disagrees with you is probably working for this enemy.

- Dismiss any alternative "solution" to your problem as "completely inadequate".

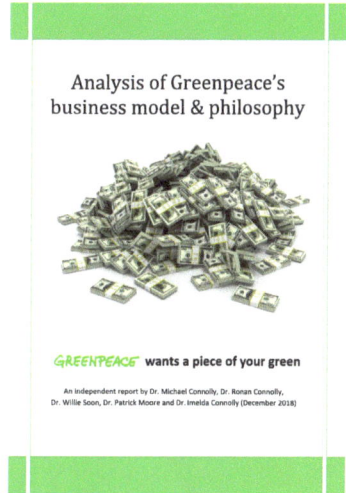

Analysis of Greenpeace's business model & philosophy

GREENPEACE wants a piece of your green

An independent report by Dr. Michael Connolly, Dr. Ronan Connolly, Dr. Willie Soon, Dr. Patrick Moore and Dr. Imelda Connolly (December 2018)

Greenpeace wants a piece of your green - An independent report by Dr. Michael Connolly, Dr. Ronan Connolly, Dr. Willie Soon, Dr. Patrick Moore and Dr. Imelda Connolly, December 2018

P. A. Moore, Confessions of a Greenpeace Dropout: The Making of a Sensible Environmentalist, Beatty Street Publishing, Inc. Canada, 2010

P. Moore, Fake Invisible Catastrophes and Threats of Doom, 2021

Dr. Moore goes on to say that other major NGOs, seeing how lucrative that strategy is, have followed suit, selling out their initial good intentions and high morals in favour of greed. Unfortunately, our media have not woken up to this shift and still treat NGOs as credible sources of information. Similarly, these NGOs have a seat at the table when governments are planning policy. This has to stop because:

"No enemy is worse than bad advice"

Sophocles

Scientists showed that the perceived moral superiority NGOs enjoy gives them a license to sin and deceive.

"This study explores why nongovernmental organizations (NGOs) engage in unethical behavior because of, and not despite, their perceived moral integrity."

"Our research reveals the dark side of moral idealization and confirms that the NGO halo effect is a risk factor for NGO unethical behavior."

"This paper is the first to establish that the NGO halo is positively related to NGO unethical behavior."

I. De Bruin Cardoso, Exploring the Dark Side of the NGO Halo: Relating NGO Mission, Morals, and People to NGO Unethical Behavior, Journal of Philanthropy, 30, e70000, 2025

NGO messaging is so strong that many plastics industry people have fallen for it and now feel ashamed of their jobs, according to internal company surveys. How can the plastics industry attract and retain top talent if this misinformation persists? It is another reason to fight back and set the record straight.

One way for NGOs to get donations is to show scary images that evoke emotions. But we now know that every image you have ever seen of a turtle with a plastic bag around its neck was made in Photoshop.

Any organisation that uses faked images to fool you into donating has given up any noble aspirations they may have once had in favour of greed and trickery. We cannot lie our way to a brighter future. We need solid data and then to make wise choices based on that information. That is the path to progress.

The problems with allowing dishonest groups to mislead the public are many. It causes a misled public to buy products that increase impact. A misled public vote for policies that make matters worse. Misled teachers teach our children misinformation. The list goes on.

This occurs via an echo chamber effect, whereby:

- They intentionally mislead the public.
- Then, they conduct a survey that reveals that the public demand action.
- Then, they demand policy to make the action a reality.

Here is an example — a recent survey collected the opinions of people on the topic of ocean plastic. These opinions will, of course, be used to call for immediate action, when in fact, none of those people have checked the science, so their opinions have no basis in reality. Readers of this book now know that the perceived "threats" mentioned here are an illusion.

B. R. Baechler et al., Public awareness and perceptions of ocean plastic pollution and support for solutions in the United States, Frontiers in Marine Science, 10, 2024

Their messaging is so strong because NGOs are packed full of marketing people, not scientists, and are often funded by billionaires.

BILLIONAIRES FUNDING FICTION

A reporter contacted me and said he felt something fishy was going on because he noticed how well-coordinated and persistent the attacks on plastic are. He asked me to keep my eyes open and alert him if I got any clues or insights into who was funding the effort. Imagine my surprise when a friend sent me a link to an article where the billionaire Michael Bloomberg openly declared that he funds the nonprofits Beyond Coal, Beyond Carbon, Beyond Petrochemicals, and now also Beyond Plastics.

E&E NEWS
By POLITICO

Publications ▾   Subscription ▾   About ▾   FREE TRIAL   LOGIN   Q

A new Washington is shaping. Stay informed with E&E News. Learn more

7-DAY UNLIMITED ACCESS   FREE TRIAL

E&ENEWSPM

Bloomberg takes on the plastics industry

By E.A. Crunden | 09/21/2022 04:29 PM EDT

Modeled after the billionaire's successful drive to shut down coal plants, the new campaign will seek to block more than 120 petrochemical projects.

Former New York City Mayor Michael Bloomberg at a conference in Idaho in 2021. Kevin Dietsch/Getty Images

https://www.eenews.net/articles/bloomberg-takes-on-the-plastics-industry/

Ironically, Michael Bloomberg's anti-plastic effort goes against his own goals. He claims to be against plastics because he believes they consume fossil fuel and increase greenhouse gas when, in fact, as we have seen, science proves the exact opposite. Plastics reduce fossil fuel consumption and greenhouse gas. This is what happens when powerful people are hoodwinked by the popular narrative and do not do their due diligence by checking the evidence. He is spending a fortune to campaign against his own stated targets. I wish I had that kind of money.

When I found out that it was Michael Bloomberg helping to fund the anti-plastics lobby, I wrote to the reporter, but interest in investigating was not forthcoming, perhaps because of the news organisation he works for. Maybe you can guess which one it is. The reporter did say that he would take a look at the CIEL and Safe Piping Matters, two organisations that raised his suspicions. What qualifies as suspicious? Simple — organisations that make claims that go against what the peer-reviewed evidence says.

When I was at INC-4 in Ottawa, I was fortunate to meet H. Fisk Johnson, CEO and Chairman of the Board at SC Johnson. We chatted, and he seemed to genuinely care about the environment, especially the oceans. I explained to him that the science shows vastly less ocean plastic than originally guessed by Jambeck and offered to show him the evidence. He said he was very interested and that his press team would contact me for a recorded video interview. Unfortunately, that never happened, and to this day, he is out there campaigning for policies that make matters worse — all because he didn't check the facts.

The lunacy is not limited to the USA. In Australia, a husband and wife who made their billions in the iron and steel industry founded the Minderoo Foundation in the name of philanthropy.

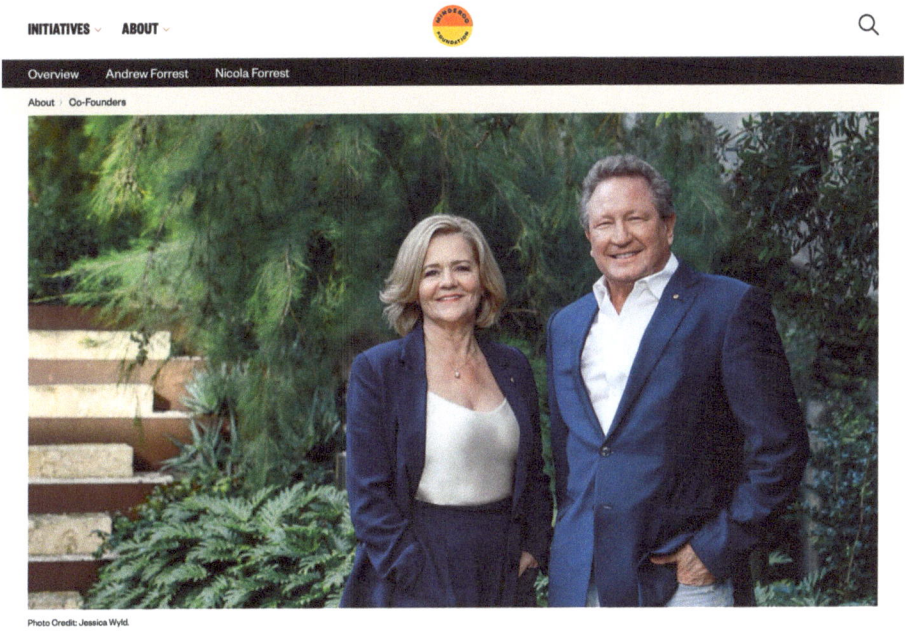

Photo Credit: Jessica Wyld.

CO-FOUNDERS

Andrew and Nicola Forrest founded Minderoo Foundation in 2001, and continue to drive its philanthropic mission today. Andrew remains Chairman of Fortescue Metals Group, the publicly listed company he founded in 2003, which is one of the world's main suppliers of iron ore. Dividends from Fortescue fund Minderoo Foundation's ongoing commitment to philanthropy. We take our name from Minderoo Station, the family homestead where Andrew grew up in the Pilbara region of Western Australia. It has been a part of the Forrest family since 1878. Minderoo is an Aboriginal word meaning permanent and clean water.

That sounds virtuous until you read the output from Minderoo, which includes unjust attacks against plastics. The science clearly shows that their iron and steel industry creates vastly more environmental impact than plastics do. Some say that the best defence is offence, so it could be handy to have your own organisation to point the finger elsewhere.

Minderoo's anti-plastics reports raised such concerns regarding accuracy that I reported it to their "Integrity Hotline" via the Deloitte whistleblower link they provide. Do you think they replied or acted? One would imagine that an organisation genuinely interested in aiding humanity would have engaged in a conversation.

Many NGOs are lying to us, and the media are only too glad to spread their message of doom because bad news sells. "What bleeds leads," as they say in the press.

NGO CREDIBILITY TEST

How can we tell which NGOs lack credibility? There are some signs to watch out for.

Firstly, NGOs where it has been announced that they are funded to attack plastics, rather than to protect the environment or respect the facts. Beyond Plastics is an obvious example. Organisations that state their desired outcome, no matter what the evidence is, are not to be trusted.

Also, look at their people. Is it a list of respected scientists or a bunch of marketing people and lobbyists? That could be a sign that they are not so interested in checking facts and are more motivated to spread some message they have been asked to share.

Compare what they say to what science says. Do they match? If an organisation consistently makes statements that are false, then that is a definite red flag.

Are they genuine enough to retract a statement if it is shown to be erroneous?

If not, then they lack integrity and cannot be trusted. The Ellen MacArthur Foundation made the infamous "More Plastic than Fish in the Ocean by 2050" statement, which has been debunked by the BBC, the CBC, and my own investigation. Did they act with honour and publish a retraction, or did they stick with the claim that suits their agenda? I have seen no retraction — have you?

The WWF told us all that we eat a credit card of plastic a week, even though that was found to be wrong, with the real number being tens of thousands of years. Did they retract their statement or are they still collecting donations based on that claim? I just checked their website, and the misinformation is still there next to a "donate" button. According to Forbes, the WWF's revenue was over $500 m. People think of NGOs as virtuous warriors that stand up to big business, but the numbers reveal they are big business, too.

Do they only show one side of the equation?

Are they living their life based on what they claim they believe or are they typing their anti-plastics mantra on a plastic keyboard, wearing polyester clothes and cellulose acetate glasses? That is a sign that they are not genuine.

A person genuinely against plastic would be in a cave on a wooden bench without electricity, a computer, internet, or a cell phone, not being paid to wage a smear campaign against the greenest choice we have. The sooner they are exposed as charlatans, the better.

When applying these simple criteria, we can make a list of some organisations with suspiciously low credibility. They include:

- Greenpeace
- WWF — World Wildlife Fund
- Sierra Club
- UNEP — United Nations Environment Programme
- WEF — World Economic Forum
- Beyond Plastics
- Break Free From Plastic
- Plastic Soup Foundation
- Ellen MacArthur Foundation
- Plastic Pollution Coalition
- Scientists' Coalition for an Effective Plastics Treaty
- Minderoo
- A Plastic Planet
- Chatham House
- SourceMaterial
- Ductile Iron Pipe Research Association (DIPRA)
- Safe Piping Matters
- CIEL — Center for International Environmental Law

Many assume that the United Nations Environmental Program (UNEP) is a reliable source of information, but when you compare their statements to the science, we see repeated and serious divergence. That includes them supporting the debunked "more plastic than fish by 2050" and other false claims about plastics related to fossil fuel and greenhouse gases. Another example is the list of 11,646 chemicals they presented that are allegedly present in plastics but that are not registered in any chemical toxicological inventory.

"UNEP's Chemicals in Plastics Report (2023) sought to document the reportedly 'often-overlooked chemical-related issues of plastic pollution, particularly their adverse impacts on human health and the environment as well as on resource efficiency and circularity'. The UNEP report was followed in March 2024 by the 'PlastChem' report, published with funding support from Norway. To highlight the abundance of information that already exists for these chemicals, ICCA compared and validated the 13,000+ chemicals identified in UNEP's Chemicals in Plastics Report to information available from global chemical inventories and toxicology information."

That headline spread quickly, but scientists checked and found that over 88% of those chemicals were registered and most with sufficient information to confirm that they are safe. So, once more UNEP made anti-plastics claims that turned out to be nowhere close to accurate.

"ICCA's analysis reveals that 88.3% (11,646) of UNEP's catalogue of 13,186 chemicals are already referenced and indexed on one or more chemical inventory."

Plastics Additives report Fact Sheet, The Global Partners for Plastics Circularity, 2024

Every industry has room for improvement, but grossly misleading the public is deeply unhelpful. The very organisations that we turn to for reliable advice have been weaponised against us. One must realise that no politically driven organisation can be expected to serve anything but their own interests. They have virtually unlimited funds and resources and thus no plausible excuse for getting the facts wrong.

UNEP also orchestrated the INC-1, INC-2, INC-3, INC-4, and INC-5 events around the world in which thousands of people flew to discuss a plastics "emergency" that NGOs invented. I calculated that the greenhouse gas emissions created from just one of those events were equivalent to 40 million PET bottles, so their events have a vast impact.

Speaking of events, people always talk about cutlery and drink containers, but a recent study confirmed what I just alluded to about events. The travel to and from the event dominates impact, not plastic knives and forks or bottles.

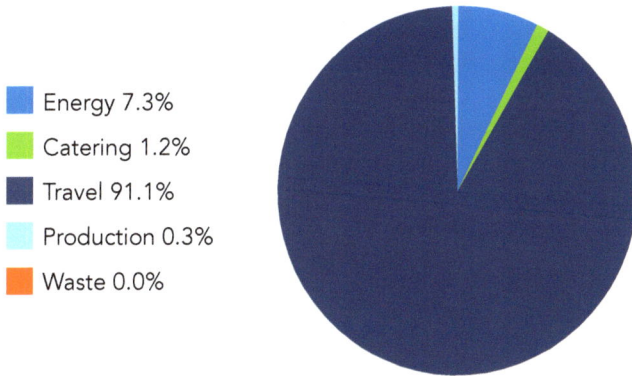

Energy 7.3%

Catering 1.2%

Travel 91.1%

Production 0.3%

Waste 0.0%

800 people created the impact of 370 cars for one year

SPC Impact 2024 Sustainability Report

THE MEDIA

The media would do well to take their job more seriously. Rather than parroting salacious nonsense, do some real investigating and break a real story, such as the one about the environmental NGOs, who have sold out and now peddle fiction for donations. That is a huge, important story, but in 5 years, I have not found one reporter who will cover it.

They should also read *The Baloney Detection Kit: Carl Sagan's Rules for Bull-shit-Busting and Critical Thinking,* in which a famous scientist guides us through the process of deciding what is true and what is not. That would help them to do their job better.

Over the last 5 years, I have written to many journalists and reporters, providing feedback supported by strong, peer-reviewed scientific evidence. The fraction of reporters who respond to facts is worryingly low. About 10% of the time, they bother to look at facts sent to them by a respected independent scientist. One time, I posted an appeal on LinkedIn asking for one true professional reporter, the kind who cares about evidence, to contact me. That post got thousands of views, but not one called me.

I found one famous investigative reporter who had won all kinds of awards. He promised to cover the story of the corrupt NGOs spreading misinformation about plastic materials for profit. He asked me to remind him, and I did once every 6 months for 2.5 years. Eventually, he showed up with a camera guy, shot hours of footage, then never aired it. And that is the state of "journalism" today.

Scientists proved the media have lied to us by grossly misrepresenting the science about microplastics.

C. Völker, J. Kramm and M. Wagner, On the Creation of Risk: Framing of Microplastics Risks in Science and Media, Global Challenges, 4 (6), 1900010, 2020

CORPORATIONS

Household name companies like Google and Amazon have announced they plan to move from plastic to paper envelopes even though every life cycle study found that this increases GHG, fossil fuel use, and waste.

FORMAT	FOSSIL FUEL CONSUMPTION (MJ-EQUIV)	GHG EMISSIONS (KG-CO$_2$-EQUIV)	WATER USE (l)	PRODUCT-TO-PACKAGE RATIO AND PERCEN WT.	PKG LANDFILLED (G)/1,000 KG MAILER
Poly Mailer	1.49	.06467	24.70	5.8:1 85.2 % : 14.8 %	166,400
Bubble Mailer	2.60 (+74.0 %)	.1092 (+68.9 %)	36.68 (+48.5 %)	3.4:1 77.1 % : 22.9 %	284,975 (+71 %)
Paper Cushion	2.34 (+56.6 %)	.3425 (+430 %)	195.68 (+692 %)	0.8:1 43.3 % : 56.7 %	972,807 (+485 %)
Paperboard	3.51 (+135 %)	.4494 (+595 %)	124.56 (+404 %)	0.7:1 41.8 % : 58.2 %	1,034,696 (+522 %)

Streamlined Life Cycle Assessment E-Commerce Mailer Packaging Case Study, Flexible Packaging Association

These companies may regret these virtue-signalling choices, assumedly designed to please their customers when those same customers realise they have been misled by the companies they trusted. In my experience, publicly traded companies often bow to such pressure to maximise shareholder profit, while privately held companies are more likely to check the facts and then do what is right, rather than what is easy. That is just my observation over recent years.

PLASTICS INDUSTRY

The plastics industry is partly to blame for letting lies go unchallenged year after year. It seems that they hoped the false accusations would go away or that their trade associations would step up and present the facts. Neither one has happened, so now we are years on, and everyone has been brainwashed with falsehoods. That is a real problem because it is much harder to change someone's mind once it is made up.

Even today, the efforts are too little, too late. I have been dismayed to see that not only are some major trade associations not showing the facts, but they have not even looked for them. Instead, they message and email me begging for help. They have had millions of dollars a year to rise to the challenge and have failed. Some of the smaller trade associations have done a much better job, but they have limited resources at their disposal.

THE PUBLIC

The public is partly to blame as well. They form strong opinions on no or bad evidence, even evidence they know is not to be trusted. They also enjoy virtue-signalling and obsessing over bags and straws because they don't want to make any sacrifice that would really help the environment and future generations.

Also, the public prefers natural products and sees plastic as synthetic. They are right that plastic is synthetic, but concrete is synthetic and so is steel. We think of paper as natural, but it is made using lots of chemicals and synthetic materials. Even wool and cotton require chemical processing and modification, such that life cycle studies find that polyester causes less impact than either cotton or wool.

Our tendency for anything that looks natural to be considered safe and good is deep within us, in our "gut," but it is a faulty way of making choices. Rattlesnake poison is natural, but that doesn't mean we should drink it.

From the public's perspective, cheap implies inferior quality and bad. So, while plastic is the least expensive and least impact-creating option for packaging, our perception of it is no doubt tainted because we throw it away every day after unwrapping whatever product it was protecting. I call it packaging amnesia. One minute we are elated that our precious cell phone or computer arrived undamaged, protected by packaging. Then, 30 seconds later, we stare at that same packaging and wonder why there's so much waste. People need to be made aware that the net effect of packaging is positive for the environment.

Many people are too busy pretending to be good. Taking a paper bag or straw is like giving a dollar to a homeless person on the street. It makes us look good and feel good, but we know it does not really help and may even make things worse. This virtue-signalling is getting in the way of progress.

PLASTIPHOBIA DIAGNOSIS

People are easily misled and reluctant to change their minds, no matter what evidence is presented. They either find an excuse not to look or ignore the evidence. Over tens of thousands of online interactions, I have seen this behaviour and the symptoms of plastiphobia again and again.

uRbaN DICTIONARY Browse ▾   Store   Blog   Discord

Q Search

Plastiphobia

The irrational and scientifically **unsubstantiated** fear of **plastics** in **the environment**. This fear has been propagated by those with monetary business interests to spread this fear in the effort to raise funds and grow their businesses by spreading falsehoods about the fanciful idea that plastics are the root if all evil in the environment.

*The not for profit companies are attempting to create Plastiphobia in an effort to raise money based on the **irrational fear** of **plastics** in **the environment**.*

by **intothefray** November 4, 2022

A person says that plastics are bad, but when you offer them a one-page summary of evidence, they will not even look at it. That means that they do not actually care about the environment.

They say that the study you showed is too old, so they can discount it. That argument makes no sense because science is valid until disproven. An experiment done a hundred years ago should give the same result today. They are just looking for a way to cling to their prejudice against plastics.

They say that the people cannot be trusted because they work for the plastics industry. That is an especially bizarre line of illogic for several reasons:

- It would imply that the only people who really understand the topic may not provide input. I wonder if when these same people are sick, they get a diagnosis from a medical professional, aka a doctor, or do they ask their car mechanic or hairdresser for their insights?

- The evidence is not my own scientific studies, anyway. Instead, I cite independent peer-reviewed science, so it is not my opinion that is in question.

- They claim that the person showing hundreds of peer-reviewed studies is "biased" when a person who has not read the science cannot possibly judge that.

- They accuse the person who cites hundreds of peer-reviewed studies of "cherry-picking" when they themselves have presented no evidence at all.

- They change the topic repeatedly, every time one of their misguided opinions based on internet myths is disproven.

- They say they are against single-use plastic but are not concerned at all about the zero-use spam that comes through our letterboxes every day and directly into the trash. They don't care about that because it's paper, and in their mind, it is good and natural, even though it is neither.

It is sad to see just how far people will go to cling to their plastiphobia, but it is somewhat understandable. Consider arachnophobia, for example. It would be hard or impossible to convince someone to stop fearing spiders.

It may help if people were to acknowledge their bias against plastics, as a first step in fighting their plastiphobia.

SCIENTISTS & POLICY

A survey found that trust in scientists is relatively high. Even so, only 50% of respondents thought scientists should be actively involved in policymaking, with the rest feeling that it is better if scientists deliver facts for others to turn into policies.

Tyson & B. Kennedy, Public Trust in Scientists and Views on Their Role in Policymaking, Pew Research Center, 2024

This gives some reason for hope. Perhaps we can present credible evidence from scientists to make some people see the light. Although, even here, the anti-plastics campaign is one step ahead, as they have their own organisations of so-called "scientists" who make bold proclamations alleging that plastic materials are a threat. These appear to be sham organisations because their "scientists" make claims counter to what the peer-reviewed science tells us.

Also, the idea that our politicians will gather evidence and then do what is best for us does not have a good track record.

VACCINATION AGAINST MISINFORMATION

The spread of misinformation has now been recognised as a real danger and even a threat to democracy.

"Concerns about misinformation are widely held, with a recent UN survey suggesting that 85% of people worldwide are worried about it."

S. van der Linden & D. R. Grimes, Misinformation really does spread like a virus, suggest mathematical models drawn from epidemiology, The Conversation, November 5th, 2024

The book Foolproof: *Why We Fall for Misinformation and How to Build Immunity* by Sander van der Linden is an excellent resource if you want to protect yourself against misinformation. It turns out that misinformation spreads like a virus; thus, the same mathematical models used in virology can be applied to combat it. It has been found that inoculation, or "prebunking" as they call it in the book, is effective at boosting our defences to prevent infection and the spread of misinformation to others. We need to teach our

children how to defend themselves, and this should become a vital part of the school curriculum. Nowadays, we have information overload, and discerning fact from fiction will be a powerful tool and a competitive edge for those who know how.

PSYCHOLOGY OF PLASTIC

At the end of my keynote talks, when people have seen convincing evidence that plastics are not to blame, one of the most common questions I get is: "Why do people dislike plastic then?" I'm not a psychologist and I could not find a scientific study, so I asked ChatGPT 4. My hopes were not high, but the answer made a lot of sense.

"What are the psychological reasons for people disliking plastic materials?

ChatGPT's reply:

The psychological reasons behind people's dislike of plastic materials are rooted in a combination of aesthetics, cultural narratives, and cognitive biases. Here's a breakdown:

1. Perceived Unnaturalness

- Artificial Appearance: Plastics are often seen as unnatural due to their synthetic origins and their glossy or artificial textures, which contrast with organic materials like wood, metal, or glass.

- Cultural Associations: Natural materials are culturally linked to sustainability, craftsmanship, and authenticity, whereas plastics are associated with mass production and disposability.

2. Durability Paradox

- Too Permanent Yet Disposable: Plastics' durability makes them environmentally problematic when discarded, yet their use in single-use items creates cognitive dissonance. This paradox triggers a sense of guilt and frustration.

- Immutability of Waste: The idea that plastics "last forever" in the environment amplifies fear and resentment, even when this perception is scientifically overstated.

3. Negative Symbolism

- Mass Production and Overconsumption: Plastics are seen as a symbol of consumerism, cheapness, and the throwaway culture.

- Pollution Iconography: Media often show plastics as the primary cause of pollution, particularly in oceans. These images evoke strong emotions, like anger and sadness, which are projected onto the material itself.

4. Aesthetic Preferences

- Lack of Warmth: Plastics often lack the tactile and visual appeal of materials like wood or stone, which are associated with warmth, elegance, and tradition.

- Perception of Fragility: Lightweight plastics can feel less substantial or "cheap," reducing their perceived value compared to heavier, denser materials.

5. Media and Misinformation

- Emotional Framing: Headlines and visuals focus on dramatic narratives about plastics (e.g., harm to wildlife), which stick in people's minds due to the availability heuristic—people judge the severity of an issue based on memorable examples.

- Confirmation Bias: Once individuals internalise negative beliefs about plastics, they unconsciously seek information that reinforces these beliefs, ignoring contradictory evidence.

6. Moral Judgments

- "Purity" and "Naturalness": In psychological terms, plastics violate notions of moral "purity" by being synthetic and seemingly disconnected from the natural world.

- Eco-Guilt: People internalise narratives that their use of plastics is personally contributing to environmental destruction, creating an emotional aversion.

7. Misplaced Blame

- Simplification of Complex Issues: Plastics become a scapegoat for broader environmental concerns, as they are visible and tangible compared to abstract contributors like fossil fuel extraction or inefficient waste management systems.

- Focus on Material, Not Behavior: It's easier to blame plastics than to address systemic issues or individual habits, like littering or overconsumption.

8. Bandwagon Effect

- Social Norms: Dislike for plastics can be amplified by social norms and peer influence. People are more likely to align with anti-plastic sentiments if these are popular or trendy within their communities.

9. Legacy of Past Innovations

- Broken Promises of Plastic: Plastics were once marketed as miraculous materials that would solve many problems, but their widespread misuse has led to disillusionment. This broken promise fosters resentment.

End quote.

Understanding these psychological drivers can help frame more effective communication and education strategies that address misconceptions and guide people toward sustainable behaviours without demonising plastics. Considering these factors, it makes sense why it has been so easy for interested parties to turn us against our greenest option, and this gives us clues about how we might fight back against the ongoing smear campaign. Hopefully, the evidence discussed in this book goes a long way to refuting those perceptions about plastic.

THE POSITIVES OF PLASTIC

Readers of *The Plastics Paradox* book and people who follow me online will know that I avoid saying anything positive about plastic. Books usually dedicate a chapter to that topic, but I have avoided the subject for various reasons. One reason is that it feels cheap to me. Often, I see plastics people say, "Look at the benefits," and it sounds to me like a desperate attempt to distract people from talking about the real issues that we face. A huge plastics industry association was in front of the US Congress to testify, and instead of presenting actual evidence, they too relied on flimsy arguments such as plastic has many benefits. Another one they tried was that the plastics industry employs many people, so please leave us alone. When I heard that argument, I couldn't help thinking to myself, "Drugs and prostitution employ a lot of people too, but that doesn't make them a great idea."

So, after 5 years of avoiding this topic, what changed? Well, a university professor who uses *The Plastics Paradox* to teach his students asked me to add something on the benefits of plastic. I asked why, and he said that if I wanted to be truly balanced, then it would only be fair to do so. I realised that he was right. After all, the paradox of plastics is that we're told that they are our greatest enemy and our best friend at the same time. To answer the paradox and work out whether plastics are a force for good or evil, we are duty-bound to look at both sides of the equation.

A major problem when it comes to plastic use is that we are not really that aware of the implications. There is an almost limitless list of items that are better and cheaper because of plastic materials spanning packaging, construction, automotive, electrical, healthcare, consumer goods, textiles, aerospace, defence, and agriculture.

While the layperson is likely to think of things like hair dryers, toothbrushes, and packaging, others will also think of medical equipment, and indeed, I have even seen some articles written by former plastiphobes who had an epiphany when plastics saved their life or the life of a loved one. Defibrillators, MRI machines, X-rays, EKGs, and all the other gear that has helped to extend our lifespan would not be possible without plastics.

But it goes way deeper than that. Modern civilisation would end overnight without plastics and so would the long-term future of humanity. That sounds a little melodramatic, but allow me to explain.

I asked ChatGPT 4 the following question:

What would happen to the electricity supply if all the plastic insulation for wires was gone?

Its reply:

> *"The absence of plastic insulation for electrical wires would lead to a near-total collapse of modern electricity-dependent systems, posing severe safety risks and causing unprecedented economic and societal disruption. Immediate adaptation would be nearly impossible, highlighting the critical role plastic insulation plays in modern infrastructure."*

Building a rocket ship and escaping the planet without plastics would also be impossible for us. As that is our only long-term strategy for the survival of the human species, it might be good for people to reflect upon the sagacity of demonising and eliminating plastics. The phrase "cut off your nose to spite your face" springs to mind.

Even the most avid anti-plastics protester has no interest in living the life they advocate for others. For example, they type furiously on a plastic keyboard, wiggling a plastic mouse, all to tell us how plastic is the work of the devil and must be eradicated.

If these people were genuinely against plastics, as they claim to be, then they would turn off the electricity to their house, throw away their computer and cell phone, and then sit in the dark to ponder the wisdom of their beliefs. At least then, we wouldn't have to listen to their nonsense anymore unless they started protesting against plastic using smoke signals and carrier pigeons! I hope they do just that and give us all a laugh.

Sometimes, when some anti-plastics nut goes too far, I tell them that if they are truly so anti-plastic, then next time they have a serious illness, be sure to mention that you want to be treated entirely without plastics. That would likely be their last decision and a win for Darwin's theory of evolution.

SUMMARY

The evidence could not be clearer — comparing vast amounts of peer-reviewed science and the story from several high profile NGOs highlights an alarming mismatch between the two. The NGOs have given us the wrong advice in every instance, and the probability of that happening by chance is zero because the scientific evidence can be found in seconds by anyone.

I can only come up with three explanations for the fact that so-called "environmental" groups are giving us advice that dramatically intensifies harm to the environment.

- Stupidity: This is statistically impossible because while any one individual may have a low IQ, NGOs have many thousands of employees, and they cannot all be morons.
- Incompetence: But these organisations manage huge campaigns that bring in billions of dollars in donations, so they cannot be called incompetent.
- Corruption: This is what the former President of Greenpeace stated, and it appears to be the only explanation that fits the evidence.

That would mean that the public and our governments are being advised by corrupt entities that have abandoned the environment in favour of slick marketing and greed. Participants at the UNEP's INC-4 event in Ottawa watched in amazement and suspicion as the NGO attendees arrived in a fleet of fancy black Escalade cars. I tried to talk to the WWF people about the scientific evidence, but they were not interested. Now, why would that be?

My comment on a LinkedIn post from A Plastic Planet was:

"This person appears to enjoy saying the opposite of what helps the environment."

That is another organisation that consistently spouts counterfactual nonsense with no disclosure about who pays them to do it.

Break Free From Plastic likes to make silly claims online, and whenever I see them, I think to myself, "Break free from the 0.5% of material that usually causes the least impact? Why?" I asked them that question, but they had no reply.

CONCLUSIONS
& SOLUTIONS

CONCLUSIONS & SOLUTIONS

Thank you for sharing this journey with me. While most people fear we are drowning in plastic and that there is no solution, we now know better. We have seen solid data distilled from thousands of studies spanning decades. The facts are known, and so are solutions that work, because once we truly understand a problem, the solution becomes obvious. Just like when we seek medical attention, suitable tests and accurate diagnosis favour a good prognosis. Here is a short recap of what we have learnt.

MATERIALS USE

We now know that materials generate around 20–25% of greenhouse gas emissions and that reducing total materials use is a positive move. Plastics make up less than 1% of materials we use, either by weight or by volume, so if we really want to make a difference, then it is time to talk about the other 99% of materials, rather than obsessing over plastics to the exclusion of all else. Plus, replacing plastic with alternatives requires 3–4 times more material and would be a large step in the wrong direction.

There is a push to limit plastic production, but as we see, that would be a counterproductive policy because replacing plastic increases materials use by fourfold.

WASTE

Waste generation mirrors materials consumption, which is logical when you think about it. Again, plastics represent under 1% of all waste and replacing them results in a 4-fold increase in waste. To illustrate this point, take your family to the kitchen and weigh a plastic bag, then a paper bag. Weigh a plastic straw, then compare it to one made of paper, metal, or glass. The results are profound and irrefutable.

Similar to the case for materials, limiting access to plastic materials or taxing them would push people to alternatives, which would result in a tremendous increase in waste.

FOSSIL FUEL

Plastics are maligned because they are made of fossil fuel, but a closer look reveals that to be an overly simplistic and misleading view. 85% of a barrel of oil is burnt, which truly is a waste of fossil fuel. In contrast, only around 5% is consumed to make plastics, which is a far wiser use of resources; also remember, at the end of life, plastic can still be burnt to recover the energy and make electricity. Not only that, but the net effect of plastics is to reduce fossil fuel use because they make cars, planes, and trucks lighter (for increased fuel economy), prevent food waste (from damage and spoilage), and insulate buildings so less energy is needed for heating. Alternative materials require far more fossil fuel to manufacture because they are more energy and resource intensive. Lastly, the majority of plastics can be made using plant-based oils instead of fossil fuel anyway, if we need to do that in the future. Such non-fossil plastic alternatives are already available at scale.

When it comes to fossil fuel, plastics production and use reduces fossil fuel use, so any action or policy that encourages a move to alternatives would be unwise and counterproductive.

GREENHOUSE GAS

Greenhouse gas (GHG) is one of the primary concerns for many. Here again, we find that the contribution of plastic has been grossly exaggerated. Plastic production creates about 3–4% of GHG, but plastic use reduces GHG by a larger amount by making vehicles lighter, preventing food waste, and insulating heat. If GHG is a concern for you, then the biggest improvements can be accomplished by driving less, flying less, and eating less meat. One return plane trip creates more GHG and uses more fossil fuel than a lifetime of PET bottles. For materials, concrete and iron/steel are by far the largest contributors, and that is where most of our efforts should be directed.

As we can see, when it comes to materials use, waste, GHG, or fossil fuel consumption, no one genuinely interested in making the world a better place would rant about plastics while completely overlooking the other 99% of the impact, and yet that is what we see today. Anyone with a genuine concern should check the evidence before deciding what to do — i.e. "check the facts before you act" — as I like to say. Acting on emotion before checking the facts often makes matters much worse, not better.

GHG mirrors the case of fossil fuel because burning fossil fuel creates carbon dioxide. The use of plastics is the best option for greenhouse gas reduction, so encouraging or forcing a move to other materials would be unjustified.

MISMANAGED WASTE: "POLLUTION" & LITTER

There is mismanaged waste in the world, but solutions are known and already in place in many countries. We know that tax on the sale of goods can be used to provide waste receptacles, collection, and proper disposal. Some countries have not yet caught up, but the pathway is clear with no special technology needed.

Scientists have discovered that what people now call plastic "pollution" is simply items that were littered in one place, then moved. So, rather than being a problem caused by companies or materials, it is a problem created by human behaviour. That is important because we have proven solutions for litter, and they are education, deposits, and fines. Blaming companies or materials for litter is unjust and counterproductive.

Litter is caused by people, and the solution to littering is called "a bin," plus encouraging people to use them.

OCEANS

Claims that the oceans are choking in plastic are based on a wild, long-disproven guess. The idea that millions of tons of plastic enter our oceans every year was simply invented, and we are told such numbers to this day, even though multiple massive studies spanning decades show measured amounts that are low and not increasing.

A sea turtle would have to swim 100,000 miles to run across a piece of plastic bag, so every image you have ever seen of a turtle with a bag around its neck is a lie created in Photoshop. How can we create a better future based on fiction and scare tactics?

Sadly, attempts at regulation completely ignore abandoned nets and other fishing gear that are scientifically proven to be what causes harm to birds, turtles, whales, and other marine life. Instead, they plan to regulate the 0.03% of ocean plastic, like bags, straws, and bottles, which are not responsible for harm. What a tragedy, and yet that is what UNEP's INC-4 and INC-5 are doing.

When it comes to the oceans, policies that would actually help are regulations on fishing nets to prevent them from being discarded, which harms marine life, as well as adjusting shipping routes to avoid whales and limiting ship speeds to reduce harm to them.

DEGRADATION

We are told that plastics don't degrade even though we see them degrade before our very eyes. There are thousands of studies spanning decades on plastic degradation. The global market for plastic stabilisers is in the billions of dollars per year. Why would anyone buy stabilisers for plastics if they really were stable? They wouldn't. Plastics degrade rapidly, more rapidly than most materials (concrete, ceramics, glass, metals) and at a similar speed to paper and wood.

Luckily, we can tune the degradation rate of plastics with those stabilisers, so a thin shopping bag contains very little stabiliser and degrades quickly outdoors. A thicker plastic pipe contains much more stabiliser and better stabilisers while providing safe, clean drinking water with an expected durability of a hundred years or more. So, the idea that plastics are bad because they don't degrade is both false and unjust discrimination.

Encouraging degradation is not a sound policy because it increases environmental impact. Durable materials tend to reduce impact. Also, degradation means converting plastic into carbon dioxide (a GHG) without capturing the energy. Burning them converts them to CO_2 too, but at least then you can use the energy to make electricity, which makes more sense. Degradable materials also increase littering. People want degradables so they can drop them on the floor and not feel guilty, and that is exactly what happens when you provide degradable alternatives.

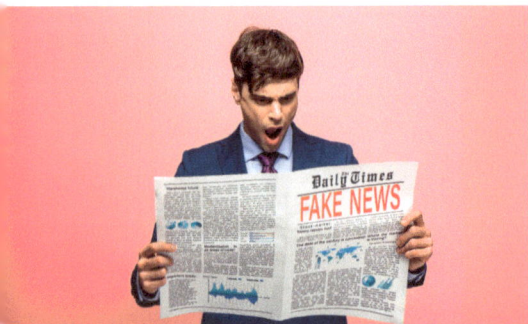

In fact, scientists found we are only concerned about microplastics in the first place because we have been misled by certain scientists and by the media. They made a big deal out of a small problem, probably to get rich and famous.

There is no new policy required here because there is already extensive regulation with ongoing testing to ensure our safety.

TOXICITY & MICROPLASTICS

Pretty much everything is toxic when the concentration is high enough. That includes oxygen, table salt, and alcohol. So, how do common plastics compare? The answer is that decades of testing show them to be some of the safest substances we have. Long-term tests show them to be safer than alcohol, table salt, caffeine, or copper, to name just a few examples.

Plastic particles, aka microplastics, are as safe as clay (i.e. dirt) or cellulose, which is what plants and trees are made of. Exposure levels are incredibly low, and most particles pass right through us. It would take tens of thousands of years to ingest just 5 g of non-toxic plastic particles. Meanwhile, we ingest 200,000 times more inorganic particles, including proven toxins and cancer-causing substances. So, while dust can pose dangers, focusing on the plastics component is a red herring.

LIFE CYCLE ANALYSIS

Life cycle analysis (LCA) is the only way to know for sure how much impact each option causes (including GHG, fossil fuel use, pollution, waste, and more). It is standardised and has been honed over decades. Even so, there remains a temptation to cheat, and thus, it is wise to check every life cycle study, rather than relying on just one or two.

Looking at hundreds of LCAs, we find that plastic is the alternative that has the least impact in over 90 % of applications studied. So, if you are not sure how to minimise your impact, then picking plastic is usually the correct choice, as proven by science.

It is not just packaging where plastic minimises impact; the same applies to water pipes, textiles, and many other use cases. One reason plastics minimise impact is that you can get the same job done using far less material, which is also why expanded polystyrene turns out to have such a low impact; after all, it is around 98 % air.

It also turns out that the least impactful choice is also the least expensive, which is great news because you can help the environment and save money at the same time. Unfortunately, at present, people are spending more on alternatives that increase impact because they have been misled by NGOs, the media, and companies looking to sell products based on false advertising and greenwashing.

Any move to limit access to our greenest choice e.g. through plastic production limits, would be unjust and harmful, resulting in vastly more materials used, waste, litter, GHG, and fossil fuel use. The same applies to taxing plastic. Taxes focused on the lowest-impact choice would just drive people to alternatives that are scientifically proven to increase harm.

RECYCLING

The perception is that we desperately need recycling to reduce the use of virgin plastic, to make plastics green, and to prevent litter. In reality, life cycle analyses show that plastics are often the lowest-impact option even with low or no recycling. Recycling is like the icing on the cake because it makes the impact of plastics even lower.

There is no correlation between recycling and litter because people choose to drop litter whether or not it can be recycled. Solutions to littering revolve around changing human behaviour, as mentioned previously.

The public is unaware that plastics recycling is well established and works well on a large proportion of common plastics like PE, PP, PET, PS, and PVC, which further reduces the energy needed by 70–80% compared to virgin (new) plastic. Mechanical recycling works and is the correct approach.

We are told by people trying to sell glass bottles or aluminium cans that we should pick that material because of a higher recycling rate, but that is a false argument. We should instead choose the material with the lowest impact then recycle that. That is the way to save the environment and save some money at the same time.

MISINFORMATION

With the advent of social media, it has never been cheaper and easier to spread misinformation. Surveys show that people have little trust in the media, and yet they have formed strong opinions about plastic based exclusively on myths from the very media they do not trust.

Today, every layperson seems to think that their opinion is as accurate as the opinions of actual experts. People who have never read a single study are happy to tell a scientist who has read thousands of studies that the scientist is wrong. Such delusional thinking is unhelpful and betrays an ego that is completely out of control. Instead, the level of conviction we have on a topic should be proportional to the amount of evidence we have to support that conviction.

Speaking of convictions, it would be helpful if we saw some charlatans at NGOs, in the media, and in the greenwashing companies fined and convicted for their deeds against society. Perhaps ensuring there is a price to pay would make them think twice.

The plastics industry needs to do vastly more to share the science, not to "defend" plastics but simply to set the record straight. They need to push their trade associations to do their job and spend their resources on this vitally important activity.

CLOSING THOUGHTS

Since I wrote *The Plastics Paradox*, so much has happened. New allegations have been made against plastics, and I have spent thousands of hours un-funded, checking the science to see whether the allegations are justified. This completely new book examines the public perception of plastic and compares perception to reality, i.e. what the peer-reviewed scientific evidence has to say.

Looking at plastics in isolation leads to incorrect conclusions. Therefore, the book takes a holistic view, including the impact of plastic materials relative to other materials and the consequences of replacing plastic with alternatives. This allows us to identify solutions proven to decrease impact and help preserve the environment.

There is a famous quote from George Bernard Shaw that goes like this:

"Two percent of the people think; three percent of the people think they think; and ninety-five percent of the people would rather die than think."

If you have read this book, then you are the 2%, and I salute you. However, that places a great responsibility on you because, with the rest of society flying on autopilot, we are the few who must make an outsized effort to preserve and protect our environment for future generations.

As we stand at the crossroads of environmental progress, we must confront an uncomfortable truth: much of what we believe about plastics is rooted in misinformation. The data is clear — plastics, when used and managed responsibly, are not the villains they have been portrayed to be. Instead, they are a vital tool in creating a sustainable future.

Imagine a world where decisions are guided by evidence, not fearmongering. Where the focus shifts from vilifying plastics to addressing the real issues — mismanaged waste, ineffective recycling systems, and the human behaviours that cause litter. This is a world in which we harness the unique advantages of plastics to reduce greenhouse gas emissions, prevent food waste, and create innovative solutions for everyday challenges.

Throughout this book, we've seen how misinformation has steered public opinion and policy in the wrong direction. Powerful entities have exploited good intentions to mislead, distract, and profit, while real solutions have been ignored. But there is hope. By embracing science and rejecting sensationalism, we can reclaim the narrative and ensure that decisions are driven by facts, not fear.

To my daughters, and to the generations that follow, I want you to know that science holds the key to progress. Truth, backed by rigorous research, has the power to dispel myths and pave the way for meaningful change. It is our duty, as stewards of this planet, to seek out that truth, challenge deceptive narratives, and make choices that benefit both humanity and the environment.

The responsibility lies with all of us. For policymakers, it means crafting regulations based on comprehensive data rather than sensational headlines. For industries, it's about continuing to innovate and prioritise sustainability. For individuals, it's a call to reject misinformation, recycle responsibly, and hold ourselves accountable for the waste we produce.

So, I leave you with this: What kind of future do we want to create? One dominated by fear and falsehoods, or one where informed decisions lead to progress and prosperity for all? The answer lies in your hands.

KEYNOTE TALKS

Bring the Visionary Behind This Book to Your Next Event. Are you looking for a keynote speaker who will challenge conventional thinking, inspire meaningful action, and deliver a message backed by scientific evidence? Dr. Chris DeArmitt, world-renowned independent plastics expert and author of *The Plastics Paradox* and Shattering the Plastics Illusion, is your ideal choice. With a track record of captivating audiences across the globe, Dr. DeArmitt brings clarity, passion, and cutting-edge insights to one of the most misunderstood topics of our time.

In his keynote talks, Dr. DeArmitt dismantles popular myths about plastics using peer-reviewed science, presenting the data that environmental NGOs and the media often ignore. His presentations are not only enlightening but also actionable, showing how individuals, organisations, and governments can make better choices for both the environment and society.

With humour, real-world anecdotes, and a passion for truth, Dr. DeArmitt captivates audiences while challenging them to think critically about the narratives they've been told. Whether addressing corporate leaders, policymakers, or educators, Dr. DeArmitt customises his presentations to meet the unique needs of your audience. Packaging, PET bottles, PS foam, plastic pipes, and microplastics are just a few potential focus areas, or request another topic to suit your circumstances.

By hiring Dr. Chris DeArmitt, you're putting your audience on the path toward a brighter future.

Let's shape a better tomorrow, one fact at a time. Book Dr. DeArmitt for your next event today!

Keynote information   **Microplastics keynote**

chris@phantomplastics.com
+1 601 620 8080

Chris is considered one of the top plastic materials scientists and problem-solvers in the world, which is why companies like Apple, P&G, LEGO, iRobot, Eaton, Total, and Disney come to him for help.

A deep understanding of materials combined with a highly creative mind allows Chris to quickly solve even the toughest challenges. To offer only one example, he solved a serious production issue that had plagued BASF for 30 years and cost them millions.

He has also received six open innovation cash prizes, placing him among the top 0.01% of innovators. In 2016, he published the book *Innovation Abyss*, which reveals the true reasons for innovation failure and the proven path to success.

In 2018, Chris was featured on CBS's 60 Minutes with Scott Pelley as an expert witness in a class-action lawsuit related to Marlex mesh plastic implants. He helped thousands of women get settlements. Later television appearances include Sky News and the BBC, as well as assorted radio and internet media interviews.

In 2020, Dr. DeArmitt published *The Plastics Paradox*, the first comprehensive, scientific overview of plastics materials and the environment covering all topics, including waste, litter, microplastics, degradation, ocean plastics, and more.

In 2024, Chris founded The Plastics Research Council, a nonprofit organisation supported by an international team of respected scientists with a mandate to provide accurate, unbiased information about plastics and the environment.

Chris has a multitude of granted patents, plus numerous articles, book chapters, encyclopaedia chapters, and conference presentations to his name. He is an award-winning keynote speaker educating global audiences on plastic materials science and dispelling myths about the environmental effects of plastics and microplastics.

www.ingramcontent.com/pod-product-compliance
Lightning Source LLC
Chambersburg PA
CBHW041307020426
42333CB00001B/2